Praise for
Love, Nature, Magic

"If you have ever wondered whether there is more to this life, this world, this universe than you can imagine, do yourself a favor and read this book. *Love, Nature, Magic* will blow your mind and open your heart. It just might help you think differently about that 'pest' insect or that 'weed' daring to crash the party of your perfectly manicured lawn. Buckle up for an exhilarating ride toward enlightenment."

—JOHN GROGAN, international bestselling author of
Marley & Me: Life and Love with the World's Worst Dog
and former editor, *Organic Gardening* magazine

"*Love, Nature, Magic* is an invitation to discover a sacred dialogue with the plants and spirits of the land, like humans had long ago before we were kicked out of Paradise. Mother Nature never stopped talking to us, as Maria Rodale discovers in her journey into the world of the shamans. It is we who stopped listening. This rich and timely book will guide you to open your heart and mind to a world infused with magic and beauty. It's all around you, really!"

—ALBERTO VILLOLDO, PhD, bestselling author of
Shaman, Healer, Sage and *One Spirit Medicine*

"Having lived and breathed organic gardening and farming her entire life—and eons before it was fashionable or cool—Maria Rodale is the best guide I can think of to explore new discoveries with. I AM ALL IN with *Love, Nature, Magic*."

—LAURIE DAVID, producer,
An Inconvenient Truth and *The Social Dilemma*;
coauthor of *Imagine It! A Handbook for a Happier Planet*

"In *Love, Nature, Magic*, Maria Rodale took me on a learning journey that I hadn't realized how much I needed to experience. With her witty prose and intimate relationship with gardening and shamanic journeying, Maria extends a loving invitation, wrapped in vivid storytelling, for readers to discover nature and the magic residing within us in beautiful and bold, new-found ways. Her book is a nurturing peace offering from nature, including messages from some of the least likely voices — poison ivy, bats, and vultures — reminding us that when we work in partnership with nature, we all benefit."

— **SHAWN DOVE**, founder,
Campaign for Black Male Achievement;
managing partner, New Profit

"In *Love, Nature, Magic*, Maria Rodale invites us to reimagine our relationship to nature, to plants, insects, and animals, and to our own souls. She invites us into a world just below the surface of our daily lives and awareness: a world that brings us closer to what is real and eternal, to what heals and what just may be the antidote to our increasing disconnection to ourselves and nature."

— **MARK HYMAN**, MD, senior advisor, Cleveland Clinic
Center for Functional Medicine; author of *Young Forever*

"Maria Rodale, in *Love, Nature, Magic*, encourages us all to reach beyond our full potential by diving into the depths of our existential selves. Then, and only then, will the dream open and the planet speak to you with the wonder and silence of creation."

— **DIANA BERESFORD-KROEGER**,
author of *To Speak for the Trees*

"Essential and stirring. Maria Rodale takes us on a journey of love, beauty, magic, and . . . mugwort! Let us heed her words and learn to live in harmony with planet Earth."

— **JOSH TICKELL** and **REBECCA HARRELL TICKELL**,
producers and directors, *Kiss the Ground* and *On Sacred Ground*

"When someone as grounded as Maria Rodale shares how the intelligence of plants, insects, and animals surpass what we have come to believe, we should listen. There is a teaching here. *Love, Nature, Magic* is about sentience, a living world more spectacular and alive than one can imagine, a shared consciousness known to Indigenous people for over ten thousand years. Once this is understood, what else can you do but honor and love all of life."

— **PAUL HAWKEN**, founder, Project Drawdown

"Maria Rodale is courageous, and so in sync. She is an instinctive adventurer in the world of natural mysteries that is derided by 'sensible' people, and her timing may be just perfect to start changing such perceptions. Maria writes that things come to us when we need them, and we sure do need love, nature, and magic right now. Or, at least, I do. I am following her on a journey that speaks to the future of interspecies harmony: Humans at peace with nature — healing, not harming. My thanks to Maria for her warmth, perspicacity, and bravery."

— **EDWINA VON GAL**, landscape designer;
founder, Perfect Earth Project

"Maria Rodale's irresistible new book is *Eat, Pray, Love* meets *Dr. Doolittle* meets, well . . . Maria Rodale. Her book wants me to ditch my phone and squish berries, smell leaves, and capture fireflies in my own backyard, not giving a damn what my neighbors think."

— **DANNY SEO**, editor-in-chief, *Naturally, Danny Seo*

"*Love, Nature, Magic* is a beautiful, heartful look at the interconnectedness of all beings, be they plant, animal, or human. Nobody is better suited to tell this story of what it means to link the sacred and the mundane than Maria Rodale. The result is an absolute must-read, a delightful road map of wisdom about how to live more softly and in sync with the natural world, rather than at battle with it."

— **ELISSA ALTMAN**, author of *Poor Man's Feast* and *Motherland*

"Our survival depends on our rediscovering that we *are* nature, not separate from it. Maria Rodale's inspiring and provocative shamanic journeys remind us how important it is to see the tapestry of life around us with the eyes of the heart. Once we see our deep connection with all life, we will not only learn how we can save it but how it will save us. You will never see nature the same way again."

— DOUGLAS ABRAMS, coauthor of
The Book of Hope and *The Book of Joy*

Dear Oliver,

LOVE
NATURE
MAGIC

Demand Organic!.
and trust the magic.

XO ♡

Maria

ALSO BY MARIA RODALE

Scratch:
Home Cooking for Everyone
Made Simple, Fun, and Totally Delicious

Organic Manifesto:
How Organic Food Can Heal Our Planet,
Feed the World, and Keep Us Safe

It's My Pleasure:
A Revolutionary Plan to Free Yourself
from Guilt and Create the Life You Want
(Maya Rodale, coauthor)

Betty's Book of Laundry Secrets
(Betty Faust, coauthor)

Maria Rodale's Organic Gardening:
Your Seasonal Companion to
Creating a Beautiful and Delicious Garden

LOVE NATURE MAGIC

shamanic journeys into
the heart of my garden

maria rodale

CHELSEA GREEN PUBLISHING
White River Junction, Vermont
London, UK

Project Manager: Angela Boyle
Editor: Fern Marshall Bradley
Copy Editor: Nancy W. Ringer
Proofreader: Deborah Heimann
Designer: Melissa Jacobson

Printed in Canada.
First printing February 2023.
10 9 8 7 6 5 4 3 2 1   23 24 25 26 27

ISBN 978-1-64502-171-1 (hardcover) | ISBN 978-1-64502-169-8 (ebook)
| ISBN 978-1-64502-170-4 (audio book)

Library of Congress Cataloging-in-Publication Data is available.

Chelsea Green Publishing
85 North Main Street, Suite 120
White River Junction, Vermont USA

Somerset House
London, UK

www.chelseagreen.com

To the shamans who have kept this wisdom alive
from the beginning of time.
Thank you.

To all the Indigenous people who have suffered.
I am sorry.

To the shamans who have worked directly with me.
Deep gratitude.

To the magical beings who taught me how to fly on my own.
I am forever grateful for your guidance.

And to my grandchildren — the born and the unborn.
I will always love you.

CONTENTS

Love

Love has a mind and heart of its own. Love cannot be forced, and yet it is a force.

Love is primal. You love some people, some things, but not others. Sometimes it hurts. Sometimes it feels overwhelming. Sometimes you wish you didn't feel it, but you do anyway. And when you don't, you can feel lonely and sad. Love is messy, raw, passionate, and desperate, sometimes even vicious.

Love is also a path. Once you step onto the path and begin to live a life based on love, a magical world opens before you. Love is one of the great mysteries of the universe. Perhaps it even is the universe.

Nature

Nature is everything. Trees, plants, soil. People, animals, insects, birds. Water, fire, air, sky, space. Buildings, cars, computers. The seen and the unseen.

Nature is what feeds us. Delights us. Inspires us. And even destroys us. We are nature. We are nothing without nature.

Nature is alive and intelligent. We are alive and intelligent. The universe is alive and intelligent. From the tiniest speck of matter, tinier even than the Higgs boson . . . to the biggest thing we can imagine . . . the whole universe and beyond — even the alleged metaverse — nature is the mystery we dream of understanding. Because when we understand nature, we just might understand ourselves and each other.

Magic

Magic is a powerful mystery. Not the sleight-of-hand trickery that entertains and often awes (although that's fun). But the kind of magic that has no logical explanation — those moments when everything seems impossibly perfect and aligned: You think of someone, and they call you. You meet someone, and they turn out to be exactly who you were meant to connect with. You hear a song that goes straight to your heart, a message from the universe reflecting just what you needed to hear.

Magic is also when you feel you are receiving messages from deceased loved ones — perhaps in the form of a feather or an animal, or the scent of someone familiar. Or when your gut tells you something is wrong, and later you find out you were correct. It's coincidence, but it's more than that. It's synchronicity, but it's more than that. It's intuition, but it's more than that. It's magic.

Trust the magic.

Welcome to My Magic Garden

D ear Reader,
I found Love, Nature, Magic in my garden. While weeding, actually. I was trying to eradicate a plant called mugwort, a seemingly pesky plant that plays a starring role in this book. Little did I know what a strange path mugwort would lead me on once I decided to try to listen to it.

Gardening (or farming, for that matter) is not for the faint of heart. In my garden I come face-to-face with birth, death, decay, murder (!) and mayhem, invasions, weeds armed with poisons and sharp weapons. After a good day in the garden, I am dirty, sweaty, scratched, and bitten and only briefly satisfied because I know that this is a relationship that will not end until I end, buried in the ground to be eaten by the decomposers and fed to the roots of trees that will outlive all of us. And notice I did not even mention the weather. Or the ticks. Or the rabbits and woodchucks.

Growing plants for food, beauty, and pleasure is a creative collaboration with the greatest artist of all — nature. It took me decades to learn that I am not the one in control, and the more I relax and pay attention to what everything in my garden wants, the happier we all are. It's not a war, it's a dance. It's not a competition, it's a partnership. It's not a race, it's a festival. It's not easy, but it's the kind of hard that builds strong muscles and creates something new and wonderful. And isn't that what we want to do in this world? Create something original and exciting, encouraging everyone to live to their full potential and enjoy this strange experience we call life? With a garden, we can create the world we want to see and live in. That, my friends, is magic.

One day, after a particularly frustrating bout in the garden trying to "control" an "invasion" of mugwort, I decided to see if I could do a shamanic journey to understand what, if anything, this plant was trying to communicate to me. (A shamanic journey is a drug-free way of going inward to seek understanding of the outer world by speaking directly with spirit in other realities. I'll explain more about journeying in a moment or two.) That experience was transformative. Fascinating. Illuminating. It made me want to journey to communicate with all the other beings in my yard that annoyed me, and that I needed or wanted to understand better.

The idea for this book came from that experience. What if, I thought, I combine my experience as a gardener, where my roots have spread and deepened and connected me to Earth, with what I have learned as a journeyer, where I have soared to other worlds? One of the things we learn when we journey is that the best way to honor the spirits who guide us is to listen and show that we heard their message. Writing a book felt like the best way I could honor them.

It was exciting to think about writing that book, but terrifying to imagine sharing it with the world. This was not the kind of thing I felt comfortable talking about with people. Except my kids, and yes, they think I'm weird. But I'm getting too old to worry about what other people think of me. The path started opening before me, and I couldn't help but follow to see where it led. The more I started talking with people about shamanic journeys, the more they shared with me about their own similar experiences and deep curiosity. It wasn't that scary after all.

We all started in a garden. Whether you believe in the biblical Garden of Eden or the theory of evolution (or both), our roots are in nature, in learning about the world around us through the plants and animals that sustain us, the rocks and trees that shelter us, and the waters that quench our bodies and souls. If we can overcome our fears and learn to see nature as a friend and ally, what is then possible? If we can transform our annoyance with plants, animals, and insects we regard as pests into love and appreciation for them, how might our whole world change? Could we transform this dystopian mental death spiral that has seduced so many people, changing it into a cycle of upward evolution and cooperation? Perhaps it's possible to course-correct our seeming environmental human plane crash by learning to love those things that frighten and annoy us.

We humans often seem trapped in a series of arguments that it feels like no one will ever win. Logic and facts don't seem to matter. We lose trust in each other and in the systems that were created to seemingly sustain us. I decided to journey because I needed to know if there was another way. I was looking for answers to satisfy my own longing to *know*, to *understand*, and to *find hope*.

Through shamanic journeying, I found those things and more.

I now believe it's possible to create a new Eden where knowledge is not a sin, desire is recognized as part of our human purpose, and love and understanding are the original blessing to be nourished and cultivated in the garden of our lives. I also believe we can create a place where diversity is celebrated and embraced as a beautiful gift from nature, because nature *thrives* on diversity.

It's easy to love the obviously delightful parts of nature, the plants and birds and other creatures that have become our friends and delight us with their beauty — the roses, the tomatoes, the lilies of the field, the butterflies, the hummingbirds. Just like it's easier to love our own families or people who look like us. But what about those beings that are harder to love? The ones we try to exterminate or eradicate? The weeds, the pests, the invasives? They may have important things to teach us. What are we capable of learning from them if we *really* listen? I realized that I stood to learn the most from the plants and animals that frustrated and frightened me. And they were eager to help teach me.

This is where my adventure begins and what these stories explore.

XO,
Maria

What Is Shamanism and Shamanic Journeying?

L et's get something clear right up front: I am not a shaman.

My friend Lisa, on the other hand, is the real thing: a gifted shaman.

I first met Lisa Weikel in 2013. I had just witnessed a friend of mine completely heal from a mental/emotional breakdown after she attended a session with Lisa. One of my friend's doctors, who very wisely involves unique healing modalities in her approach, had referred her to Lisa. My friend's healing was so transformative (and her story to tell, not mine) that I summoned up the courage to go see Lisa myself. As a then-CEO of a health and wellness publisher, I felt it was my duty, as well as my passion, to explore the frontiers of healing (as safely as possible).

I had no idea what to expect, but Lisa looked normal — short gray wavy hair, comfortable clothes and shoes, wearing a bit of spiritual-looking jewelry. She drove a Prius. I learned she's a mom of three sons, and she has a sly sense of humor, a devoted husband, and a variety of Boston terriers and cats. She's also a lawyer. I felt safe with her.

According to Lisa, shamans are energetic healers of the human spirit "who travel into other realms of consciousness in order to work with spirits on behalf of individuals or their communities." She explains, "Historically, shamans were usually either born to the work, displaying certain characteristics, abilities, and 'gifts' that were recognized as such within their culture, or were essentially conscripted into it via a near-death experience. Many were mentored — and still are, in Indigenous cultures especially — by other shamans within their community. This process is a bit different in cultures like ours that have, until around the mid-twentieth century, dismissed shamanic states of consciousness as ecstasy or even psychosis."

Lisa was introduced to shamanism and shamanic journeying in the early 1990s by anthropologist Michael Harner, the author of a classic book called *The Way of the Shaman*. After journeying herself for ten years, she was "unexpectedly initiated and set upon the path of shamanic healing by the Tuvan shaman Aldyn-Herel Choodular, while in Siberia, at the direction, and with the permission, of Professor Mongush B. Kenin-Lopsan."* (That's also Lisa's story to tell.)

She was then trained in more practical healing methods by the Four Winds Society, which was founded by Dr. Alberto Villoldo, a medical anthropologist and author of many books on shamanism.

But I didn't know all this back then. All I knew was that I was lying on a blanket on the floor of Lisa's private office in Bucks County, Pennsylvania, about to be "worked on." The session lasted only a few hours, but it was nothing short of miraculous for me. Turns out I had a lot of healing to do — from a recent divorce, a challenging family history, and a few other traumas collected over lifetimes. She introduced me to my power animals. And she gave me a new life contract authored by my very own soul, with Lisa being the scribe. She told me to burn the old one. I did.

I'm not going to tell you what my power animal is (or what my life contract says either). These are obviously very personal and not what this book is about. Power animals (also called spirit animals or totems) and spirit guides are the way nature and spirit reach out to

---

* Professor Kenin-Lopsan was named a Living Treasure of Shamanism by the Foundation for Shamanic Studies for his heroic work in preserving shamanism in the Republic of Tuva during the Soviet era, when the government attempted to eradicate Indigenous practices.

us to connect, guide, and protect us, regardless of our race, religion, gender, nationality, ethnicity, or sexual orientation. Spirit guides may appear as animals, angels, ancestors, or other mythical beings. According to Lisa, most of us are born with one or two power animals or allies who guide and protect us throughout our lifetime (guardian angels, if you will), while others come and go as needed. A power animal is a very special relationship, built over time, on trust, intimacy, and confidentiality. There is no "reflected glory" benefit from having a certain type of power animal. The power comes from the relationship, which is up to each of us to cultivate and honor. If we use the relationship properly, the power grows and grows. If we abuse the relationship or use the power for selfish or harmful purposes, the power may be withdrawn. In that way, it's like any true friendship.

Here's something very important about shamanism: Just because someone says they are a shaman (ahem, like the "QAnon organic shaman" guy who stormed the US Capitol on January 6, 2021), or even has studied as a shaman, that doesn't make them a good shaman. Power is power. Some use it wisely. Others use it for less than noble purposes. Shamans can work for good or for ill, but the primary attribute of shamans is their ability to travel to other realms of reality to retrieve information or power on behalf of individuals or communities, usually in service of healing, finding lost things, or working with nature. Lisa doesn't refer to herself as a shaman; she prefers the term *shamanic practitioner*. I still call her a shaman, though. She uses her power with pure intent and is humble and discreet about her work, which involves healing or otherwise being of service to others.

Going to see a shaman like Lisa, or even Dr. Villoldo, doesn't require shamanic journeying on your part, just like going to a doctor doesn't require you to be a doctor yourself or to follow through on all the instructions the doctor gives you. And just as you don't have to be a monk to meditate, or a yogi to practice yoga, you don't have to be a shaman to do shamanic journeying. At my first meeting with Lisa, we didn't talk about journeying. What I experienced during our session felt like what I now know as a shamanic journey. But at the time, I had no idea what journeying was or how to do it. All I knew was that during that session, in my mind's eye I saw deeply moving and bizarre things that gave me a new perspective on my life. And because I had experienced a vivid dream about my power animal even before this session, Lisa naming it validated the entire experience for me.

Here's how Lisa explains it: "During a session, I actually journey on your behalf and work with your allies to work with your soul to, among other things, shift its patterns, and — if directed to do so — retrieve a power animal to help you integrate the work. You do not journey during our session. You relax and enter into a meditative state. Almost everyone I work on, when in that relaxed state, 'sees' stuff. But you do not journey, as you do not consciously travel to another realm with intent."

All I knew was that after the session ended, I felt a deep sense of healing and was curious to know more.

A few months after I saw Lisa for the first time, I went to a spa for a break from the stresses of life and my job. Out of a sense of quiet desperation about my future mixed with my constant drive to do research, I made an appointment with a psychic. (I look at psychics the way I look at everything. It's all research.) She told me that I needed to go to the redwoods — what I was searching for could be found there. The redwoods comment went into my mental file folder on psychic exploration, and I went home.

The day I returned to the office, I found an invitation to speak at the EcoFarm conference at Asilomar, a retreat center on the Monterey Peninsula in California. The conference would be at the time of my birthday in early 2014. I immediately knew I wanted to go because my father had also spoken at Asilomar, and I remembered him talking about it with a sparkle in his eye. Plus, animal behaviorist and autism activist Temple Grandin would be presenting at the conference, and I really wanted to hear her speak. Additionally, I had always wanted to check out Esalen, the iconic spiritual education center in Big Sur, about ninety minutes' drive south of Asilomar (coincidentally, Esalen happens to be near a redwood forest).

To be completely honest, as a recently divorced woman, I assumed the psychic meant that I would find a boyfriend in the redwoods. Instead, I found shamanic journeying. On the weekend I would be there, the Esalen schedule offered only one workshop that caught my interest — a spiritual massage and shamanism session with the Hopi-trained shaman Bo Montenegro and a Brazilian spiritual healer whose name was Maria Lucia Bittencourt Sauer. Look, this is how the universe works. Maria is my name. Lucia is my youngest daughter's name. Shamanism is interesting to me. I signed up. I later found out that Maria Lucia had a daughter named Maya who is the exact same age as my oldest daughter, Maya.

At Esalen, for the first time I took part in a guided journeying session while someone was drumming. After Bo spoke a special set of words to "open sacred space," as the practice is known, we all reclined in a circle on the floor, made comfortable with pillows and blankets. The lights went out, the blinds were drawn, blocking the incredible views of the Pacific Ocean, and Bo softly, steadily beat the drum. As I journeyed, I was gobsmacked by the clarity and beauty of what unfolded inside my psyche — visual experiences filled with animals and symbols. I felt myself flying and receiving strange messages. When the drumming stopped, we all shared our journeys. Each person's journey was completely different and all of them were spectacular and magical. I was hooked. I wanted to learn everything about journeying.

I left Esalen excited but also disoriented. Reality as I had known it shifted — now I felt like I had found a door to a secret universe. As a birthday gift to myself I drove up Route 1 to stay one night at the Post Ranch Inn, which had been on my hotel bucket list because of its eco-conscious reputation and spectacular architecture perched high above the Pacific Ocean. The inn offered sessions with a shaman, but I hadn't had time to sign up. Instead, in the gift shop I bought the book *Dreaming Your World into Being* by Jon Rasmussen, who was the house shaman at the inn. The universe was giving me clues that this was a path I was invited to walk on. I had no idea where it would lead, but I would follow wherever it went because it was so fascinating.

When I got home, I downloaded a drumming app simply called Drum Journey, by Mindful Bear Apps, and started journeying even more on my own. I read everything that showed up on my Trail of Books, which is what I call the magic of finding just the right book, film, or documentary exactly when you need it most. What I learned was like finding the missing pieces to a puzzle I'd been working on my whole life, trying to understand those things that unite us rather than divide us.

In almost every culture around the world there is some tradition of shamanism and shamanic journeying. These ancient Indigenous traditions of seeking knowledge, insight, and communication with spirit guides in other realms involve shifting your consciousness, often through the use of a rhythmic sound, usually a drum or a rattle, although the shift can also be achieved through dance or the use of other sounds.

Dr. Michael Harner may be the person who is most responsible for bringing shamanism and shamanic journeying into contemporary Western (white) consciousness. Dr. Harner identified what he called "core shamanism," the shamanic practices that are similar all over the world.

The word *shaman* comes from the Siberian word *saman*, which means "one who knows" or "one who sees." But every culture has its own word for the important role a shaman plays. In Peru the p'aqos of the Q'ero people, high in the Andes, access other realms on behalf of their community and are charged with responsibilities from healing to acting as intermediaries with elemental forces. Mexico has curanderos/curanderas and the nagual tradition of the Toltec, among others (Don Miguel Ruiz, author of *The Four Agreements*, comes from this tradition). In North America many Indigenous people use the English term *medicine people*. In Australia the Aboriginal people have healers called ngangkari. The Nordic countries have a female shamanic tradition called seidr, and the reindeer-herding Sami, also from the Nordic countries, have their own powerful shamanic traditions. The Dagara culture of West Africa calls their shamans boburo; the Zulu of South Africa call them sangoma. There is even a Netflix anime series, *Shaman King*, that deals with Asian shamanism. Basically, there are as many names for shamans as there are cultures in the world.

But often shamans have been persecuted and killed. In North Korea, shamanism is still banned, and shamans can be executed because they represent the past and not the modern future. "This was true in Siberia as well," Lisa explained to me. "When the Soviet Union annexed Tuva in the 1940s, for instance, there was a concerted effort to exterminate all shamans." In Europe and the British Isles beginning in the fifteenth century, many female shamanic medicine healers were accused of witchcraft and murdered.

When it comes to shamanic *journeying*, there are many methods in different cultures for traveling to other worlds. In the Middle East, Sufi spinning is a method to transport people to other realms. In Japan, the mystical practice of Shintoism works with nature spirits. In India and parts of Asia, meditation is the doorway into direct experience with the divine. In Europe and Scandinavia, there are pre-Christian, nature-based traditions of inner exploration and traveling to other worlds, including the Celtic traditions of Ireland, Scotland, Wales, and parts of England. Many African cultures use drum and dance to transport people, with each practicing its own unique tradition. In South and Central America, drumming and dance are powerful traditions of spiritual transportation. (Journeying is different than meditation, however. Meditation is primarily a process of quieting the mind, whereas journeying is a method of activating a different part of the mind to visit other realms.)

This is by no means a comprehensive list of names, places, and types of journeying, or an in-depth explanation.

Sadly, much history of shamanism has been lost due to religions and politicians trying to eliminate the individualistic power of shamans. But as I learned about the wide range of shamanic traditions, I (and others, like Dr. Harner before me) became convinced that there is something universal about shamanism. In all these traditions, drumming and rattling and the ringing of bells and singing and dancing play a part. Throughout history, drums were banned in many places because they are such a powerful tool for direct spiritual experiences, which can be a threat to the established order.

Some cultures use medicines derived from plants, mushrooms, and even animals as part of shamanic experience, working with everything from ayahuasca (a vining plant native to South America) to psilocybin mushrooms, peyote, mescaline (from the San Pedro cactus), secretions from the cane toad, and tobacco (the pure form, without any additives to make it addictive). It seems the desire to visit inner and other worlds is universal.

However, there's an important point to understand here: Using psychoactive plants or drugs *is not necessary* to delve into the inner world of our psyches and connect with spirit to discover insights and truths about ourselves and our world. With or without the use of entheogens (often called plant medicines), *sound* is the universal highway to spiritual journeying. The sound of a drum or a rattle, especially. I do *not* use drugs to journey. I have never tried ayahuasca because I really, really don't like to throw up. As a mother of three daughters and the owner and CEO of a third-generation family business that employed hundreds of people, I didn't have the time or interest to travel into the heart of the Amazon and risk getting sick or worse. The great news is that I discovered I didn't need take any hallucinogens or embark on gonzo adventures to visit other realms. All the journeys I describe in this book were done without drugs (unless you count coffee) and in the comfort of my own home.

Global shamanic practices are part of the worldview (cosmology) known as Animism, which is the belief that everything has a soul and is alive and is connected. Or Pantheism, which is the belief that God is the Universe, and the Universe is God.

Shamanism, if respected properly, can peacefully coexist with any religion precisely because it is *not* a religion.

Studying religious history is one of my lifelong hobbies. I have come to believe that all religions arose from local cultures trying to make sense of, honor, and control the world around them and help people find and stay on their path. Many religions are tribal and focused on maintaining a cohesive group identity. Others are more focused on empowering the individual to find their own way. Religion has served an important role in human evolution, and so have Indigenous spiritual practices. Again, shamanism is not a religion, it is a *practice*. I am not trying to convert anyone. There is no group to join. No dogma. No leader. No guru. No rules. You don't even need to believe in God. You can profess any religion you want and still appreciate and work with shamanic journeying or consult with a shaman. And just like any practice, the more you do it, the easier it gets.

The more I have learned about shamanism, the more I am impressed by the academic credibility of the people who have studied it. These are not people who wanted to become gurus or invent some New Age, pseudoscientific fantasy. They are scholars who are genuinely curious about understanding Indigenous healing, and they worked closely and respectfully with various local experts and came to understand that shamanism is a key method to help people heal, find wisdom, and develop a direct connection with spirit. Whether that spirit is outside us or inside us may not really matter. What matters is that shamanic practices really help people heal and deepen their understanding of themselves and the world around them. It's important to note that Dr. Harner and others worked with the willing encouragement of Indigenous shamans to translate these practices for the benefit of all. After all, there is a species to save. (Humans.)

In the first peer-reviewed scientific study of its kind, published in the journal *Frontiers in Human Neuroscience* in 2021, neurologists from the University of Michigan and other institutions looked at shamans' state of mind while at rest, during drumming sessions, and while listening to classical music, comparing them against a control group and against a second group of subjects under the influence of psychedelics. They assessed their subjects using EEG recordings (a medical technology that records brain activity) and questionnaires. The researchers found that shamans *do* enter an altered state of consciousness when listening to the drumming, noting, "Shamanic practitioners were significantly different from controls in several domains of altered states of consciousness, with scores comparable to or exceeding that of healthy volunteers under the influence of psychedelics."[*] In other words, something really *is* happening when you close your eyes and start the drumming. If scientists and researchers are ready to take shamanism seriously, there is a wealth of important research yet to be done.

My goal is not to teach you how to journey, but over the course of this book, I do describe some of the basic practice of journeying and how I approach my journeys. I include some of my key sources in "The Trail of Books" sources on page 217 if you are interested in doing it yourself. Whether you ever decide to try journeying is up to you. I wrote this book simply to tell you some unusual and fascinating stories and to share what I've learned through journeying. As you read, I hope it will open up your heart and mind to new ways of thinking — especially about our relationship with nature. All the journeys I describe took place at my home, surrounded by my garden, with plants, animals, insects, one snake, and one bird that I had personal — often difficult — relationships with. I have a much better understanding of all of them now, and I feel a deep love for them. Even the *most* annoying ones!

When I journey, I frequently experience a conversation between my logical mind, which asks whether this is really real or am I making it all up, and my mythical mind, which immerses me in the magic of the journey to crazy and glorious places where I see things that seem impossible. This mixed state of belief and disbelief is common for many people as they journey, especially in the beginning. But

---

[*] Emma R. Huels, Hyoungkyu Kim, UnCheol Lee, Tarik Bel-Bahar, Angelo V. Colmenero, Amanda Nelson, Stefanie Blain-Moraes, George A. Mashour, and Richard E. Harris, "Neural Correlates of the Shamanic State of Consciousness" *Frontiers in Human Neuroscience* 15 (2021), https://doi.org/10.3389/fnhum.2021.610466.

as two shamans told me in response to my questioning about this dichotomy: "Does it really matter?"

No. It doesn't matter. Because whatever it is that happens during a journey, I learn and discover things I wouldn't see otherwise. Whether it's simply a deep dive into my own psyche or an actual trip through the galaxies, does it matter? No. It's all about the journey. My journey.

When I first began this whole adventure, I had a lot of anxieties. Would bad things happen if I said the words to open sacred space wrong? What if I can't remember the words? (Let's face it, I can't.) Is it OK to use sage smudge sticks from the supermarket? What if I start opening sacred space in the east instead of the north or south? Am I "culturally appropriating" things that don't belong to me if I do this stuff? Will some mean person storm into my kitchen wearing a nun's outfit with a ruler in her hand if I don't do things perfectly? Ouch!

After much research, practice, and questioning I realized there are only a few things that are truly important. If you ever do decide to journey, here is what is most essential: First, have the *right intention*. If your heart is in the right place, you don't need to say all the exact words. Second, *respect*. You can use your own cultural tools and herbs, or you can learn from other traditions, as long as you are showing respect — respect for other cultures, respect for nature, and respect for the process. And always ask for permission. Third, *relax*. Tell your inner mean person (or your inner Mother, Father, Spouse, Sibling, or whoever seems to disapprove of whatever you do) to get lost. This is *your* practice. *Your* time to explore your inner worlds. *Your* journey. There is no right or wrong and no perfect. Although please, try to travel with love in your heart. Fourth and finally, *gratitude*. Always express your thanks for whatever guidance you receive.

Shamanic journeying is all about having a direct experience with the mystery of the universe. *Your* universe. *My* universe. They are both valid. You can call it whatever you want — God, Consciousness, Energy, Source, Science, the Universe. I call it Love, Nature, Magic.

Getting Grounded

B efore I begin telling my stories, I want to set the scene and share a bit about my unique background so that we are standing together on common ground as you learn about the other realms I visit.

These journeys took place in a comfortable spot in my house that is private and close to the outdoors and my garden. I love to garden. And especially to landscape. It's where I can express my creativity most freely. My garden is an oasis in the woods on top of a very small mountain in eastern Pennsylvania (elevation 850 feet). Sixty years ago, the property was a farm, but it had been abandoned and taken over by brambles and vines. At the center of the land, down a dirt road bordered by giant rocks, ferns, moss, and wild trees, is a place called the Sand Pit. It is an old sand mine near the peak of the mountain, and it looks like a tiny desert. The first time I saw it, I fell in love.

When my then-husband and I found this beautiful piece of land nearly twenty years ago, we knew we wanted to build an environmentally sound and beautiful home surrounded by an organic garden on one of the most brambly acres. The local architect we worked with told us he was a shaman (this will matter later in the book), but that was long before I even knew what a shaman was. At that time, I was fully caught up in life as an executive, a mom, a wife, and a daughter. I was vice chairman of the Rodale family publishing company. Maya was off at college, and my second daughter, Eve, was still at home. (Lucia wasn't born yet; she came along a few years later.) My mother was the company's chairman, and she was suffering from cancer. In other words, I had a lot going on.

I would be starting the landscape from scratch, a blank canvas that included several large sandy dirt piles and lots of rocks. And my vision was not just to plant a vegetable patch. (Because of the Rodale Institute's well-known research work on organic farming systems, a lot of people think I am a farmer. I am not a farmer.) No, I would be designing a whole new world.

Because I still wasn't confident in my own design abilities, I briefly hired a local landscaper who specialized in native plants. But when I told him I wanted to create a magic fairy-tale garden, he looked at me like I was nuts. I thanked him for his plan and never looked at it again.

I had studied permaculture with Bill Mollison, the eccentric Australian who based his theory of landscape design on working with Indigenous cultures in Australia and Africa. So I knew the first step in the design process was just to sit and observe. Where was I in the land around me, in the mountain range, in the country, in the universe? Which way was north, south, east, and west? How did the water flow? What did the land seem to want? What did all of us want?

From observation and listening, a landscape plan emerged in my mind, and I'm happy to tell you that over the years it has (mostly) come to fruition. I'm almost out of space to plant anything new. I grow vegetables, herbs, and flowers. Trees have reached their mature height, and bushes have filled out, their arms spread in joy. There's an orchard and beehives. I've got chickens and a stumpery. And I've got stories to tell.

I've also gotten to know the land better. I learned that before it was a farm it was heavily harvested for wood. Before that, the Leni Lenape lived on this rocky (very rocky) mountaintop of what is known as the Pennsylvania Highlands, which are part of the foothills of the Appalachian Mountains. The Leni Lenape (whose name translates to "real, original person") are matrilineal and matrilocal. The removal of the Lenape from this region wasn't sudden and violent. It happened over a period of four hundred years, a combination of disease, failed negotiations with European settlers, fraud, and bad deals including the Walking Purchase deception of the 1730s. But even after that, many Lenape fought side by side with the Continental Army during the Revolutionary War and remain in Pennsylvania today. Bethlehem, the city nearest to where I live, was founded by the Moravians, who welcomed Indigenous people into their communities. I do not know the exact Indigenous history of the land where I now live and garden, but I try to honor the Lenape legacy.

In the times before there were any humans at all living on this land, there was a glacier. How do I know this? Because it basically stopped right in the middle of my property and melted, leaving behind . . . the Sand Pit! A geology professor from a local university brings his class every September to study the unique layering of time recorded in the sand. His annual email to me is one of the first warnings that fall is coming, which always makes me a little sad because summer is my favorite season.

Gardening and landscaping are my art but also my solace. The shamanic term *miqui*, "mulching," refers to working with the earth and asking it to help you release your struggles and turn pain and heavy energies into mulch, which gardeners know is a good thing that feeds new life and keeps weeds from taking over.

I've done a lot of mulching. After my father died in a car accident when he was only sixty years old, I grieved by planting trees. As my mother struggled with cancer and died, I planted perennials. While my marriage was falling apart, I planted an orchard. During the times I was a stressed-out mother, I dragged my daughters along on my frequent trips in search of unusual plants and herbs and let them pick out plants too. I celebrated becoming CEO of Rodale, Inc., by building a spectacular composting area. I grew vegetables and fruits and preserved them while watching the whole publishing industry implode. Wherever I traveled — and I traveled a lot for both business and pleasure — I brought home planting ideas, garden ornaments, rocks, and shells to remind me of how beautiful the world is. When I sold the family business, I cooked from my garden to soothe and nourish my family and myself. During the Covid pandemic, I decided to raise chickens so we would have a steady supply of fresh eggs. (I don't eat my chickens because I love them too much. But I do eat chickens I don't know personally.) And when all my children and my granddaughter came to live with me for the first six months of lockdown, we all planted, built more gardens, harvested, and were grateful to be safe from the chaos of the world.

I sometimes have struggled to reconcile all the different parts of my life and the varied roles I have had to perform. But it turns out being a crazy gardener and a CEO (albeit a former one) are not that different. Both involve strategic thinking, long-term planning, vision, management of a diverse and often unruly group of characters, dogged persistence (aka stubbornness), and listening and observation. (All these skills are crucial to being a mother, as well.) Shamanic journeying added yet another point of connection, because through business, gardening, and shamanism I learned to create the world I want to see and how to make dreams come true (what some call manifesting). At this moment in time that's something we all need to learn how to do. Our lives depend on it.

I was excited to realize that shamanic journeying is a tool for speaking directly with my garden and all of nature. I wanted to learn how everyone can engage with nature in a way that is collaborative rather than combative. I wanted to know how working with nature can teach us to be better humans. And I *needed* to learn how to open my own heart more to the magic.

One thing I have learned as a lifelong gardener is that life is not really about survival of the fittest. It's not even about adaptation. (Sorry Darwin.) It's really about survival of the happiest. The plants and creatures that are happiest are the ones that thrive. And each plant (or animal, for that matter) requires different things to be happy. Some need sun, others shade. Some prefer dry, others wet. But if they are in the right spot and encouraged with love, they become happy and thrive. They spread to fill in the empty spots. They come back again and again every year. They surprise us by popping up in new places nearby. They don't need fertilizer or chemicals. They are just happy, and they know it. I don't need to know exactly why this is so, because I believe in the magic of it all. It's the magic of regeneration, which we all have access to.

I watch my eldest granddaughter play in my garden for hours, talking to the fairies, telling stories to herself, finding treasures, singing. She vehemently believes in magic. She is happy (most of the time).

Do we lose our happiness when we stop believing in magic? Or are she and I happy because we have dirt, which is loaded with serotonin, under our gnarly nails? Does it matter?

My older brother died of AIDS in the 1980s. At one time, he lived in a little cottage in the woods, and nearby there was a patch of anemones that thrived, even without attention. I dug up some of them to plant in my own garden. They have spread beautifully and gently. They are happy and they know it. And they remind me how precious time is, and how easy things can be when we pay attention to what nature really wants. We just have to be willing to see what's right in front of us, in all its simple beauty.

We often tend to look for magic in unusual things — faraway places, gourmet foods, high-risk entertainments, riches, fame, or approval from others. But really, magic is all around us wherever we are. I began experiencing it most when I was finally forced by the pandemic to stay home. When I became still, it became easier to hear what nature was trying to tell me and to feel the love all around me. I found that magic is everywhere, and it is powerful. It took me a long time to find it. But now that I have, I'm excited to share what I've learned.

All the journeys in this book except one (Vulture) happened in the order in which they appear. No names have been changed and everything is true.

There is one more thing we need to do before we dive into the journeys together, and that is to open sacred space . . .

Opening Sacred Space

Before starting a journey, it's important to open sacred space — which creates an area of protection around us to keep us safe while traveling — to make sure that the power we interact with is positive and good. While I do not literally take you on a shamanic journey in this book, I want you all to know the beautiful power of opening sacred space so you can understand the reverence with which people journey. I would also like to ensure that whatever you read in the rest of this book brings only good things to you!

I love hearing Lisa Weikel open sacred space because her words are so beautiful and relaxing. She combines the traditional Four Winds Society version with her own elaborations.

When I journey alone, my opening of sacred space is shorter and to the point (like me). Sometimes I listen to a recording of others opening sacred space rather than speaking words myself. Remember, what matters most is heartfelt intention. But let's start my journey stories with Lisa's luxurious opening words, so you have a sense of what it's like when I say, "She opened sacred space."

To the Winds of the North,
Saywarkintey, Royal Hummingbird, Huanakari,
To the Grandmothers and Grandfathers who've come before us
and all those who will come after us, our children's children's children:
Come! Be here with us now as we do this work.
Ancestors and Ancient Ones, warm your hands by our fire, whisper
your wisdom in our ears.
Hummingbird, help us be still amidst the chaos of the outside world so
we can drink deeply of the sweetness and wisdom that surrounds us.
Wolf, leader of the Clan of Teachers, please go out and gather our
greatest teachers and ask them to sit in council with us, helping us
remember and embrace the lessons we set for ourselves in this lifetime.

To the Winds of the East,
Apuchine, Hatun Kuntur, Huaman, Hatun Tuku: Eagle, (Great)
Condor, Hawk, and (Great) Owl:
Come! Be here with us now as we do this work.
Teach us how to fly wing-to-wing with Great Spirit, high among the
Apus, the Sacred Mountains, so we can look down on our lives and
our destinies and see our experiences from a different perspective.
Help us see the choices before us and pick up the strand of destiny
that calls to our heart most brilliantly.
Please join us, Ancient Ones — our personal, sacred mountains — as
well as all of the Appalachians, the Andes, the Rockies, the Hima-
layas, Uluru, Mount Fuji, Mount Kilimanjaro, the Caucasus,
the Pyrenees, the Scottish Highlands and Irish Cliffs, and all the
mountains from all over the world.
Please protect our medicine space and bring us your accumulated
wisdom.

To the Winds of the South,
Sachamama, Great Serpent, Amaru:
Come! Be here with us now as we do this work.
Wrap your coils of golden light around us.

*Help us shed our fears and worries, our old beliefs and prejudices,
    or whatever may weigh us down or hold us back from being the
    brightest expressions of our true selves. Help us shed these, just as
    you shed your skin.*

*Remind us to walk gently upon Mother Earth, sensitive to and aware
    of our connection to Her.*

*Porcupine! Little one, please join us as well. Bring us your qualities of
    trust and faith, innocence and playfulness, all the while remind-
    ing us that we have the ability (and responsibility) to set and keep
    effective and appropriate boundaries.*

To the Winds of the West,

*Otorongo, Mother/Sister Jaguar, Choquachinchay, Rainbow Jaguar,
    you who makes the great journey from this world to the next, and
    back again; Grandmother and Grandfather Bear, keepers of the
    wisdom that resides within all of us:*

Come and join us in this work.

*Guard our medicine space as we journey into the dark unknown,
    discovering and exploring what we may have hidden from our-
    selves; give us the eyes to see and the ears to hear that which will
    aid us in understanding ourselves and our purpose here on Earth.
    You have seen the birth and death of universes, so come and help
    us die to the old ways of being that no longer serve us.*

Remind us that, with you by our side, we have nothing to fear.

*Above all, help us approach this work and live our lives with the
    highest integrity and impeccability.*

Pachamama, Danu, Gaia, Sweet Mother Earth:

You are the Mother who never leaves us.

*You feed us, clothe us, and shelter us. You provide us with everything
    we need not only to live but to thrive. We honor you and ask you to
    join with us, consciously, not only to heal ourselves, but to heal all
    your children, two-legged, four-legged, many-legged, no-legged,
    winged ones, furred ones, finned ones, stone people, plant people,*

*standing tall people. We offer the healing benefits of our work
to All Our Relations to use as they are willing and able. Aho
mitakuye oasin.*

*Spirits of the Land where we live and where we are at this moment:
Thank you for all you do to keep our homes and spaces in balance.
Thank you for inviting us to reside, work, and play here. Please,
take your place by our fire and allow the work done and insight
achieved within this Sacred Space to cleanse, benefit, and heal
you — and this land — as well.*

***Intitayta, Mama Kia, Hatun Chaskas (Father Sun, Grand-
mother Moon, Brothers and Sisters of the Great Star Nations),
Illatixi, Wiracocha, Wakan Tanka, Ascended Masters, Angelic
Realm, God, Goddess, All That Is, You Who Are Known by a
Thousand Names, and You Who Are the Unnameable:***
Please come, shine Your light down upon us,
*Illuminate our hearts, minds, bodies, and fill us with your wisdom,
your insight, your unconditional love and compassion. Help us
feel your forgiveness and peace, as well as joy and gratitude for
being allowed to live one more day in beauty and grace.*

AND NOW WE BEGIN

*These are my musings and journeys
on love, nature, and magic.*

Bedtime stories for sweet dreams.

Journeys into the mysteries of life.

*Seeds planted in the soul of the universe
to make new things grow.*

After all, we are all nature. Nature is us.

*When we begin to understand nature,
we begin to understand ourselves.*

And that's a special form of magic.

I trust it.

I am a journeyer.

I am a gardener.

These are my stories.

Mugwort

Plants rule the world.

JOURNEY: APRIL 8, 2021

It all started with mugwort.

Its pungent leaves look like chrysanthemums, sage green with jagged edges. If it is allowed to grow to its full height of about three feet, it produces tiny nondescript flowers.

"What is this?" I asked the fancy landscaper who was visiting my garden.

"Ugh, that's mugwort. It's almost impossible to get rid of without Roundup."

Damn. I am definitely *not* a Roundup user. But that summer I considered it. Which for me was sacrilegious! After all, I belong to the family that started the modern organic movement in America, and I've written a book called *Organic Manifesto*. I searched for other options and discovered that, yes, mugwort is almost impossible to eradicate. I imagined how easy it would be to just spray it away. Sigh.

Nope. No Roundup for me. I would do it the hard way.

In pursuit of mugwort eradication, I dug down deep into the soil, pulling up roots that ran like frustratingly tangled computer cords from one plant to another, diving under big rocks and out the other side. I collected piles and piles of mugwort. I occasionally found what seemed to be a mother root that connected all the other roots, and with deep, dark satisfaction I pulled out these giant spider legs of the dread pirate mugwort.

As a lifelong passionate gardener and landscaper, I had been seduced and inspired by gorgeous photos in gardening magazines, scenes where the wildness had been meticulously manicured and cultivated. The only weed was the occasional mullein, looking fabulous in bloom, like a statuesque movie star in a bright yellow dress, standing out from the crowd. I was charmed by those country retreats of the European landed gentry even though I *knew from experience* that such photographs were taken only after an army of gardeners and stylists had swarmed in and worked for hours to create that effortlessly fabulous look of subtle luxury. In those photos, a perfectly laid table of food fresh from the garden gleamed in the golden light of sunset, sparkling like a bubbly glass of rosé wine, seducing the viewer into an ideal of relaxed sophistication. Every linen perfectly ironed. Every basket handmade. Everything under control and pleasing to the eye. These gardens weren't just beautiful fantasies — they represented a longing to fit in to a social stratum I knew I would never belong to. For decades I was determined to strive for this ideal: a garden without mugwort and other nasty weeds.

Perhaps my wish to be part of the high society of gardening harkened back to my own grandparents' desires to fit in. My father's father, J. I. Rodale, was born and raised in the Jewish tenement buildings of the Lower East Side of Manhattan. My father's mother, Anna, was a coal miner's daughter from the dismal mining town of Tamaqua, Pennsylvania. She moved to New York City when she was orphaned at the age of fourteen, and that's where my grandparents met (at a dime-a-dance hall). When J. I. and Anna bought a farm in Allentown, Pennsylvania, a few years after they married, they were eager to try out all the ideas they had learned about gardening without chemicals from the British aristocracy. They read about composting in Sir Albert Howard's famous book, *An Agricultural Testament*. And they read about the importance of healthy soil in Lady Eve Balfour's *The Living Soil*. In the 1940s and '50s, my grandparents created a formal and functional organic farm that looked more European than American. They had created their own vision of the American

Dream. This magical farm was where my family lived and where I grew up, free to roam my whole childhood.

With that as my family legacy, my garden had to live up to high expectations — both my own and what I believed others expected. And that's how I found myself determined to eradicate the mugwort. It was the first summer of the Covid pandemic, and I made the "perfect garden" campaign my primary focus. During that enforced time at home I dug so hard and so long that for one week I felt like I had won. But it was only early summer. I should have known how the story would go. After all, I'd been weeding for my entire life of fifty-nine years (if I count my early mud-pie-making days). By fall I gave up. A brief bout of Lyme disease in September led me to take the kind of antibiotics that cause sun sensitivity, and I stayed inside. Winter came. It snowed in feet, not inches, in December, January, and February.

In April I went out to my garden and there was the mugwort . . . bigger and better than ever, rising like Jesus from the dead. It was like I hadn't even tried to get rid of it the year before. I started to stab the cold earth with my Japanese weeding knife, still determined to eradicate it.

After a few minutes in stern attack mode, I stood up, took a breath, and thought for a moment. When I was a CEO and staff members complained to me about the same thing over and over again, I would say to them, "If you feel like you are banging your head against a brick wall, maybe it's time to try something different." I looked down at the single mugwort plant I had been about to murder. Maybe it was time for me to take my own management advice. The leaves were kind of pretty. I suddenly sensed the plant trying to get my attention, and I felt something shift in me — as if a different lens was placed over my eyes, with a direct line to my soul. For a brief second, I felt the aliveness — the consciousness — of everything. Especially that little sprout of mugwort. What was I doing? I was expending all this energy and fury, and it was *just a plant*. Who said it was bad? Was this how I wanted to spend my life? Was this the hill I wanted to die on? Why was I making mugwort my garden nemesis?

I consider myself a nature lover, and yet that summer, when it came to my garden, I was acting like a fascist dictator. Not only was it not a good look, but it wasn't bringing me any joy or satisfaction. Something wasn't right. What wasn't I seeing that I needed to see?

There are times when you know in your brain what's right or smart, but your lifetime conditioning and habits override your good sense, and you keep on doing something that's stupid and doesn't work. Memories rose up of my mother kneeling in her flower garden angrily pulling the weeds. I remembered her marching around our farm and complaining that the gardens weren't weed-free enough. She was Pennsylvania Dutch, and I later attributed her abhorrence for weeds to her German sense of obsessive tidiness. Even though she was long dead, I still often heard her voice of judgment in my head, saying: "Clean up this mess!"

Another voice joined in: "Make sure you pull the weed out by the roots." It was my first farm boss, sternly giving me instructions for my summer job on the farm. He was also Pennsylvania Dutch, and he pronounced roots more like "ruts." That phrase had been echoing in my mind for forty-five years.

That moment in my garden with mugwort triggered a major awakening in me. I saw how my convoluted history with gardens and my family and the Rodale publishing company was like the tangled mugwort roots. But I was finally free from the burden of the family business, having sold it two years before, and I suddenly felt able to look at everything with a new perspective — not the one I had been born with and trained for in my whole life so far, but with a new openness. Beginner's mind, as it is sometimes called. An idea began to take shape. I had been journeying for years — usually to try and solve relationship or business problems — but now it occurred to me that perhaps during a journey I could try to speak directly with a plant or animal.

Coincidentally (or magically), that very night I planned to attend a group journeying session. Lisa had started weekly sessions to provide people with a sense of community during the pandemic. She calls the practice of people gathering to journey on a regular basis a "shamanic caravan," with each of the individuals journeying in their own "wagon." Kind of like a tribe of travelers traversing a new and challenging landscape together. We were a small group, usually two to six people, who showed up to journey over Zoom with the goal of harnessing the power of group energy and intention while encouraging each person to tap into their own guidance. Lisa would open sacred space and rattle while we journeyed. And then we would share our journeys with each other.

Experienced journeyers recommend that you always set an intention before you journey. Ask a question, perhaps. Or decide in advance where you might want to go or who you want to speak with. Otherwise you might get lost or meet up with some bad influences. For many years I didn't know that. Now I do.

That night at the group journeying session, I set the intention of trying to understand mugwort and see if it had anything to tell me. The sky was still light, but I knew it would turn to dark blue by the time the journey was done. I lay down on the blanket I use for journeying — I call it my magic blanket just for fun. (It's a rainbow-colored Beaver State Pendleton blanket I bought in Santa Fe on my twenty-first birthday.) I like to lie down fairly flat when I journey. I've heard the process described as creating a "hollow reed" — an empty conduit for the energy of the journey to travel through unimpeded. I make sure my hair is loose and my shoes are off and I get comfortable.

Lisa opened sacred space. Then the rattling started.

* * *

Every journey starts with entering an imaginary portal: a hole in a tree, a cave, a body of water, or some natural crack in the earth. For this journey, I attempted to begin by entering into mugwort. It wouldn't let me in. Hmmm. I had never experienced this problem before. (During the discussion afterward, Lisa explained that living things aren't portals for journeying. She wasn't surprised that I couldn't get in through the actual plant. I had experienced a core shamanic truth before I had learned that there was such a truth.)

I changed my approach and entered my usual hole in a tree. It was dark, and it stayed dark. I felt like I was underground. I waited. When I'm on a journey and nothing is happening, I usually start to question my sanity. I started questioning it now. I waited some more. Suddenly she started talking softly to me.

"You are always trying to kill me. Have you ever thought about asking me nicely?"

No, I hadn't.

"You know you are never going to get rid of us. Plants rule the world."

That's true, I thought. Why hadn't I seen that before?

"Mugwort is good for the heart, you know. Do your research. Whenever a plant shows up as a pest, study it. We are just trying to communicate with you."

OK, OK, I got it. I had been worried about my heart because there is a history of heart attacks in my family. Also, recently I had been yawning a lot and not able to stop, which I worried was a yet undiscovered risk factor for heart-related issues.

At that moment, my faithful spirit animal showed up and said, "Let's go for a walk." We started walking in my garden, among the mugwort. "You know, all your anger at weeds is not good for your heart. Stop worrying so much about them."

The rattle sped up faster, calling me back.

✳ ✳ ✳

Was it really that simple? Just stop worrying? Easier said than done, right?

It's easy to think we are rational beings. That what we have been taught is always true and facts are facts. But through journeying I suddenly *understand* with my whole heart and soul. Mugwort is not my enemy, she is my ally. My teacher. And the more I try to kill her, the more I am eradicating any opportunity for me to learn and grow. And by the way, I *can't* destroy her because plants rule the world.

Once I started paying attention in a new way, mugwort showed up everywhere.

The very next night Lucia asked me to watch *Spirited Away*, a classic Japanese animated movie, with her. Halfway through the film, the bath master gives someone a ticket for a mugwort bath. I sit up. Huh?!

Clearly, I needed to learn more about mugwort. I had looked it up in some herb books, and I knew that mugwort is considered a healing plant. It's recommended to enhance dreaming and deep sleep, but I'm a good sleeper. I didn't need help with that. There had to be something more . . . I sensed a mystery that needed solving.

I am not a systematic or scientific researcher, but I am a voracious reader. I *love* to learn. I think I actually "suffer" from epistemophilia, which is the sometimes excessive love of and thirst for knowledge. I devour all sorts of content. For me, the miracle of modern technology is the wealth of information to be found at my fingertips online. The joy that comes from connecting the dots between history, science, nature, and the humanities is one of my greatest life pleasures.

I discovered a world of mugwort lore, products, and treatments. There is mugwort for sale online. Wait a minute . . . People buy mugwort? Apparently in Asia mugwort is a considered a very healing plant. It's often used to reduce skin redness. I often have skin redness. It's also used in acupuncture in a weird burning technique called moxibustion. Tiny piles of dried mugwort are placed on the skin at key meridian points (Chinese medicine energetic spots) and then burned. It doesn't hurt. That triggers a memory of a treatment I once had at a spa in Texas. (I'll try *almost* anything once.)

Mugwort is considered a key part of the Korean creation myth. The story goes that a tiger and a bear named Ungnyeo lived together in a cave and prayed to their divine king to be made into humans. The king gave them twenty cloves of garlic and a bundle of mugwort and ordered them to stay in the cave for one hundred days, eating only the garlic and mugwort. The tiger left after twenty days, but the bear remained. And on the twenty-first day the divine king turned her into a woman. Ungnyeo prayed to a sacred birch tree for a child, and the king took mercy on her and gave her a son, who became the founder of the Korean nation. Clearly, mugwort has some royal roots.

Mugwort's Latin name is *Artemisia vulgaris*. It's considered a plant of protection — even protection from insects and evil spirits. In many European pagan traditions, mugwort is used for smudging, similar to the way Indigenous Americans use sage or cedar. (Actually, some Indigenous Americans use mugwort too.)

Smudging is a way of cleansing the energy around a person or a thing, and like shamanism, smudging rituals are found all over the world. It is believed that the smoke gets rid of negative energy — but also that smoke can send messages to the ancestors or spirits. It's why burning incense is a traditional practice in many Asian cultures. Smudging usually involves burning a bundle of dried herbs and letting the smoke surround a person, a room, or a place that is in need of protecting, cleansing, healing, blessing, or connecting to the spirit world. Australian Aboriginal people do it. Scottish people burn juniper in a tradition called saining. Catholic priests do it when they burn incense in a thurible as they begin a mass. And believe it or not, the Pennsylvania Dutch do it too . . . with mugwort!

Now I realize that Mugwort might have been asking me: "Don't you know who *I am*??" Uhh, no. Apparently I didn't.

Here I had a sacred, powerful medicine plant in my own garden, and I had been trying to destroy it. Stupid me.

The next day I picked a bunch of mugwort, steeped it in hot water, strained it, and added the tea to my hot bathwater. That night I did have intense dreams.

I started asking my friends who are interested in the healing qualities of plants about mugwort and if and how they use it. Some people make smudge sticks from it. Others eat it. A long-buried memory surfaced of eating mugwort rice balls at Miya's Sushi (Chef Bun Lai's former restaurant in New Haven, Connecticut, that celebrated eating invasive species). They were yummy. Mugwort has a distinct and strong herby flavor that is hard to forget. I began to feel intense gratitude for this pervasive and magical plant.

I had embarked on a journey to talk with Mugwort because I wanted to understand plants and nature in a new way. To listen rather than destroy. To ask rather than demand. To learn to respect rather than disregard. And also, to find out why it annoyed me so much. But what I found was much more than some interesting information about a seemingly problematic plant. Mugwort had a message for me about a different species altogether: *Homo sapiens*. We humans are never going to get rid of our "enemies" — whatever group of humans or pests that annoy us or we don't like. All the genocides, all the wars, all the revolutions, and all the terrorist attacks throughout history have not eradicated whatever we fear is the enemy. Killing one dictator does not eradicate dictators. Trying to exterminate or "cleanse" an ethnic or religious group is not only abhorrent but doesn't work and never will. The root systems of our species and all our ethnicities are deep and strong. Trying to eliminate a group, just like trying to eliminate a plant, is a futile endeavor. Sure, lots of things have gone extinct over the eons for many different reasons. But by intentionally trying to make something else extinct, we are probably most likely to extinct ourselves.

There is a phenomenon known to farmers who plant genetically modified crops (GMOs, or bioengineered plants) that can withstand exposure to the herbicide Roundup wherein the more Roundup they apply, the larger and more resilient the weeds eventually become. The weeds know how to evolve to withstand the herbicide. They are called "superweeds." Farmers have to use even more toxic herbicides to kill those superweeds. But in the process, these farmers are harming the soil, the water, and the health of all species on Earth even more than before. This doesn't mean farmers (or homeowners, for that matter) who use Roundup are bad people. It just means they've fallen for a false story. The latest superweed for farmers is a plant called Palmer amaranth. It grows eight feet tall, with stalks two to three inches thick. Each seed head produces up to 500,000 seeds. Because competition from Palmer amaranth can reduce crop yields by up to 90 percent, conventional farmers are having to abandon fields populated by Palmer amaranth, since there are no longer any herbicides that will kill the superweed. Is the answer to develop even more toxic herbicides? Mugwort was telling me no: This is a war humans can never win. Plants rule the world.

Mugwort was also inviting me to understand that while humans must stop trying to wipe out "invasive" plants, we also need to stop hating and trying to eradicate groups of our own species — especially those that are not like us. What history has taught us, and is still teaching us, is that we are all capable of becoming fascists. When information is limited and our self-awareness muted, it's easy to lose our sense of right and wrong. In search of a moral code, we often look to words written in ancient texts. What if, instead, we looked to nature? Nature shows us that everything has a purpose, and diversity is essential. Therefore, it behooves us to appreciate our difference and listen to other perspectives. Learn from them. Befriend them. Work together with them. That is truly the only way we can find peace with one another.

After my journey, I approached the mugwort in my garden completely differently. I asked nicely. I pulled gently. I picked some for smudging. I tried putting some in soup (it was good). I let a lot of it grow wherever it wanted to. *I softened my heart.* I thanked it. And lo and behold, the area of my garden where mugwort had bothered me the most cleared up easily. And the rest stayed happily contained in a few out-of-the-way spots.

What if we change our perspective from feeling the need to control nature to enjoying and appreciating the surprises and gifts? When wild plants show up in our gardens, can we resist the impulse to eradicate them just because we didn't buy them in a pot from the local plant store? Yes. We can.

And what if that which we need to heal us (and maybe even save us) isn't something far away, outside of ourselves, or packaged in plastic in the pharmacy? What if we don't need a high-tech convoluted solution? Maybe the healing we need most is waiting to be discovered right in our own backyards. Maybe even by the side of the road.

Wild. Powerful. Free.

Thank you, Mugwort.

Multiflora Rose

That which hurts can also heal.

JOURNEY: MAY 13, 2021

It was Mother's Day, and my gift to myself was to dig up and remove all the wild plants that had sprouted up in the area inside my abandoned chicken coop. The coop, which once housed guinea hens, was perfect for raising birds, with wire mesh fencing on all four sides and even the roof to prevent attacks by predators from above. There were two outdoor coops. On the side with chickens, the hens had happily kept the coop clear. But on the other side were wild wineberries, multiflora rose, poison ivy, Oriental bittersweet, and a few other "invasive weeds." (Yes, I know that Mugwort taught me I need to question what's invasive, but these plants were really in the wrong spot.)

My plan, once the coop was rescued from the invaders, was to turn it into a berry patch — with blueberries and my favorite black raspberries — that would be protected from wild birds eating the best before I could get to them. I could imagine how beautiful it would be. But first, there were lots of plants to remove.

The Oriental bittersweet vines had grown up the outside of the fence, tightly entwining themselves through the wire mesh and covering half the ceiling with quite lovely leaves. There was also plenty of multiflora rose. I decided that, sadly, they had to go, or they would shade my berries too much. I snipped and pulled and untangled.

WHACK! One of the bigger Oriental bittersweet vines snapped back and punched me in the mouth. I brought my sweatshirt up to my mouth. There was blood. I went inside and cleaned the wound. While I was there, I looked up Oriental bittersweet and discovered that the plant has a long list of healing benefits — everything from resolving skin troubles and ulcers to shrinking tumors. Scrolling down the list on my computer screen, I said, "OK, I hear you. I'm sorry."

Back outside, I considered changing my plan, but there was plenty of bittersweet growing all over the woods around my yard. I snipped a few more branches and then moved into the coop to remove the last remaining sharp and tangled mess. A giant multiflora rose was the last holdout. The multifloras outside the coop had come out easily. But this, I could tell, was the mother of multifloras. The root went down deep, and I fought that root like it was my Moby Dick. No dice. So I apologized, and I asked nicely. I dug deeper and deeper. Still the root would not release. Finally I said: If you come out nicely, I will write a chapter about you in my book. I hadn't even yet decided to write this book, but I was ready to try anything to get that bugger out.

Pop! Out it came. So, here is the chapter on multiflora rose . . .

One of the first interesting things I learned about this plant is it is hermaphroditic — both male and female. And yes, this species of rose is considered invasive. But like many things modern humans consider invasive, multiflora was intentionally transported across continents to help solve some problems. It was brought to North America from Asia in 1850 as a rootstock for other species of roses. (Most modern roses you buy in a store are actually grafted onto a sturdy rootstock. That's why if the top part of a rose bush dies, a gnarly wild rose will often grow up in its place instead. I have learned to buy only "root grown" roses, which are much more resilient.) Multiflora was also brought in to control erosion and create natural fences around live-stock pastures. (This will be a repeating theme in this book . . . most invasive plants were brought to new places on purpose by humans who thought they were helping.) Birds enjoy eating the berries of multiflora, and the thorny thickets protect wildlife. But as in many situations where humans decide to control nature, nature takes con-trol and creates new challenges.

An article about *Rosa multiflora* published by Penn State Extension states:

> Like other shrubs with attractive flowers, **multiflora rose persists in our landscape partly due to citizen unwillingness to remove plants perceived to have aesthetic value or value to pollinators and other wildlife**. However, the dense, monocultural thickets created by multiflora rose degrade natural environments [sic] and reduce native plant and wildlife diversity.

I bolded the part that I find the most interesting. It embodies the philosophy of almost every conservationist, environmentalist, and landscaper I've encountered. Sure, it's a pretty plant and the birds, animals, and bees (and even some humans) like it — but it doesn't *belong* here, so we need to kill it! (Even though we brought it here in the first place. Oops.)

The timing was perfect for another Thursday-night Zoom journey with Lisa and the caravan. After receiving such great insight from speaking with Mugwort, I decided to set my intention to understand the multiflora rose, but I was also upset by news of Palestinians and Israel at war again. *Again.* Sigh. So that was on my mind as well.

✳ ✳ ✳

The rattling started. I entered into the tree and it led me to a path. I walked up to a cottage in the woods that I was familiar with from previous journeys, but this time, instead of being warm and welcoming, it was dark and covered in thorns . . . multiflora rose. My great-great-grandmother, a medicine woman from Lithuania, whom I know only as "Granny," walked out from behind the cottage.

"See!" she said angrily. "This is what happens when you don't take care of things!" (I come from a long line of angry women. On both sides.)

We sat down on a rock and talked about things. Family things. Tragedies set into motion by her anger at the doctor who bled her

daughter — my great-grandmother — to death (back when "bleeding" was a prescribed treatment for just about everything). Her anger
at my grandfather's family for forcing my Catholic grandmother,
Anna, to convert to Judaism. A curse she put on him and his offspring:
My father, my brother, both dead too soon. "Can you forgive me?"
she asked. I honestly had already done so (thanks to Lisa's help), and
I said, "Yes, I forgive you." But then, thinking of my grief over the
deaths of the men in my family, I started to cry.

My grandfather's death is both notorious and tragic. J. I. Rodale
died suddenly of a heart attack while he was a guest on *The Dick
Cavett Show*. And my brother, considered the heir to the business,
died of AIDS three days after he was diagnosed, when he was only
thirty years old. It was December 1985, in the very early days of that
epidemic, and my family was devastated. Five years later, my father,
Robert Rodale, was killed in a car accident in Moscow. (He was in
Russia to help establish a Russian-language organic gardening magazine.) The entire Rodale family and the family business were thrown
into chaos. My grandmother Anna (Granny's granddaughter)
outlived them all, dying after a long bout of dementia at the age of
ninety-five. She never recovered from the shock of my father's death.

"Here," Granny said, handing me cup of tea in a delicate china
teacup. I looked in the cup and saw a multiflora rose petal floating
in the pale golden tea. "It is good for heartbreak and anger," she told
me. As I drank it, the vines imprisoning the cottage receded, and it
suddenly came back to life. Warm glowing lights shone from within.
The sun started to shine again. Flowers bloomed. Birds sang. All was
well. Granny started to wash my feet in a bowl of water, but then she
made a cut in the tops of both my feet. Blood seeped into the bowl
of water, and she threw the water onto the ground. It surprised and
shocked me and I wondered why she had done that. Then a beautiful Black woman named Eve (whom I had encountered in previous
journeys) emerged from the cottage and said, "It is woman's blood
that heals the earth and creates things. Men fight because they are
jealous of our power. They think spilling their blood will create

things, but it doesn't." She paused. "Men are stupid," she said, and she spat on the ground.

✳ ✳ ✳

The journey ended abruptly. I sat up and began writing it all down. Journeys are a bit like vivid dreams — if you don't write them down, they may disappear into the void. I also find that going back and reading my notes later helps me understand the full message of the journeys. Sometimes they don't make sense until years later.

I was shocked and surprised by the anger toward men. But then I remember that the multiflora rose is a hermaphrodite. Both male and female. I've often thought that everyone is both a little bit male and a little bit female. We are all capable of hurting and healing, destroying and creating, no matter what our gender. (In fact, I have since found out that Granny was a sniper when the Russian Cossacks invaded Lithuania. The Russians won. So she and her daughter fled to America.)

Are men *really* stupid? Sometimes. Sometimes women are stupid too. Here is what is extremely stupid — fighting and spilling blood through violence. I wonder: Is the high rate of suicide among veterans due in part to the trauma of witnessing firsthand how ineffective and cruel war is? Especially when the war is unjust, as we have seen too often. In our society we often romanticize war. Kids like to play at it. Grown-ups like to make and go see movies about it. In the past it was a way for men not only to demonstrate their honor and heroism but to find comradery with other men, work, and advancement and wealth. And it still is. War is a big business. But the more we all can watch wars unfold on social media and the news, the more clearly we can recognize the terrible waste and horror. War is a very bad business to be in, and a really stupid way to resolve conflict. We bomb people's homes and kill civilians, destroying their hospitals, schools, and villages, and then are surprised and upset when immigration and terrorism become a problem.

Yes, there are circumstances when we need to protect ourselves (and especially our children). Thorns help with that. But that which has thorns can also, sometimes, heal. Multiflora rose is still a *rose*. And a cup of multiflora rose tea is high in vitamins C and E.

A cup of tea and its nutrients will not bring back my grandfather, father, and brother from the dead. And my tears won't either. But I believe that the work I do to heal my own heart will carry forward to benefit my children and their children and their children. I will use my blood to create something new. Something good. For the benefit of everyone.

My wild and weedy chicken coop is now the thriving berry patch I imagined. There are blueberries and black raspberries and an old wooden chair to sit on while snacking on berries or having a cup of tea. The woods around me are filled with multiflora rose (and Oriental bittersweet), and there they can happily stay, feeding the birds and bees and protecting all the wild things, including me.

Thank you, Multiflora Rose.

Vulture

You are not your body.

FIRST JOURNEY: AROUND 2014
SECOND JOURNEY: JANUARY 25, 2022

Vultures and I have a thing going. On the top of this small mountain where I live, there always seems to be plenty of them. At first glance they might seem scary or menacing, with their "ugly" red heads and dirty job of cleaning up dead things. But I love them. Because without vultures the world would smell like dead things, and that's a really awful smell. But I wonder if there is more to it than that. I wonder if vultures also clean up dead spiritual energy that we release during times of stress, change, or emotional trauma.

This possibility first struck me many years ago, when my husband and I were sitting on our outside porch couch, having the talk that our marriage was over. The pain was palpable. His, for being rejected. Mine, for hurting our family. Ours, for realizing what would never be in our future — anniversaries, old age companionship. In the middle of the conversation dozens of vultures suddenly started circling overhead — circling and squawking so loudly we had to stop talking. I remember thinking that they had gathered to clean the carcass of our marriage. At that time I had just begun my quest into understanding the spiritual side of nature, but I took note.

A few years later I started journeying, which was all good except for once. There has been only one time I brought myself out of a journey early because I was scared. Yes, you can stop a journey if you want to. You are always in control. You simply retrace your steps and come back out wherever you entered. I have also since learned that I can tell something to go away or run away from it. (Although I didn't know this back then, so I literally just opened my eyes and sat up.)

* * *

In this journey, I found myself in a dark, barren place that was filled with bones and dry tan dust. Vultures were everywhere, pooping their white poop all over everything (their poop, I later learned, sterilizes their feet, which is helpful because their feet come in contact with so much dead and decaying stuff). I was already feeling frightened, and then a giant masked dragon man, covered in inky black scales and with giant glossy black wings, rose up out of the ground before me. That's exactly the point when I said, "Oh shit, get me out of here," and I opened my eyes and sat up.

* * *

Afterward, the image of the winged black dragon man haunted me. I searched for information about this figure online, and I found it was like Quetzalcoatl, the feathered serpent god of the Aztecs. Or what the Mayan called Kukulkan. Or it could have just been a big dragon man. Whatever it might have been, it scared the hell out of me.

At the time I didn't understand the journey at all. But looking back from this present moment, it makes more sense. I had been struggling to try to figure out how to save the family business, but every effort led to dead ends.

In 2009, shortly before my mother died of cancer, I became the CEO of Rodale, Inc. What had started as a small family business in the 1940s had become an international publishing company (we published *Men's Health* magazine in ninety-nine countries). But in the twenty-first century, the potential of websites and search engines was the sparkly new thing for everyone, offering entrepreneurs fantasies of becoming billionaires, advertisers unlimited targeting and reach, and customers unlimited content . . . *for free*. Rodale, Inc., had to compete with brash internet startups that had no historical infrastructure baggage, no profitability constraints (thanks to opportunistic tech investors focused primarily on exit strategies), and no pensions to fund. Even in the face of those challenges, my goal was to be a good leader for the company and then pass the company on to the next generation.

Yet I was starting to suspect that the topics Rodale magazines and health books focused on (and made the most profit on) — weight loss, diets, intense exercising, and medical advice supported by pharmaceutical advertising — weren't all they were cracked up to be. And while I loved our gardening books and magazines, people weren't buying them anymore. What seemed most true to me was that living and eating *in moderation*, combined with a healthy dose of pleasure, joy, and love, were what led to a long and productive life, not obsessively tracking what you did or did not eat, how much you exercised, or how thin you were. I was in search of the answer to "What, if anything, is really right?" How can I (and our customers) live a long, healthy, joyful life, making the best contribution to humanity while we are here?

The employees kept asking me what my "vision" for the future was. I did all the research. I talked with people inside and outside the company. I analyzed all our customer and market data. And then I went into the woods to meditate and seek my vision. The problem was that the future I envisioned didn't fit the DNA of the company. And through my research, it became very clear to me that if humans didn't change their behaviors as a species, the environment would likely become uninhabitable for humankind sooner rather than later. Nature would be fine; it's us who would become extinct. And more advertising or six-pack abs weren't going to solve any of those problems.

Everyone at Rodale was frustrated and scared about what was happening in the business, and a lot of people blamed me. Learning to protect myself and my energy became a priority. This is what I was up against when Vulture showed up in my journey, and I wasn't ready to face whatever it was trying to show me.

Years later, after the difficult but unanimous decision had been made (at the urging of our next generation) to sell the business to a much larger publishing company, I came to understand that shamanic encounter with the "dragon man" as a foreshadowing. The dragon man was a symbol of death and of resurrection. But he was also a symbol of the power I had — if and when I chose to wield it.

Vulture history is filled with mixed meanings, with these birds being associated with everything from evil to holiness. They are almost always associated with death. In fact, some cultures conduct "sky burials," leaving corpses out on the tops of mountains for the vultures to devour. It is believed that the birds will take the souls of the departed up to heaven — and also a sensible way to dispose of corpses in places where burial or cremation are difficult or too expensive. Our fear of vultures reflects our fear of death. But they are harmless, really, and just doing their work. And they won't hurt you or kill you. They will only eat you if you are already dead.

One day I was hiking in my woods with a man who bow-hunts deer there. He took me off the beaten path to show me where he wanted to put up a tree stand. We came to a huge rock formation, and sitting on top of it, like some scene from a scary movie, was a "wake" of vultures. I realized then that I am providing a good home to them, and I felt grateful. I honored them and smiled. They do important work in the ecosystem. We have become friends.

And now this vulture story turns magical . . .

In the summer of 2021, I asked Lisa for a meet up to discuss some questions I had about my future, and writing this book in particular. (Should I? Or shouldn't I?) I had spent much of the last three years letting go of the past and figuring out who I wanted to be now that I was free from my life in the world of business. I felt ready to begin something totally new, and I wanted her guidance and insight on the way forward. The mugwort journey had planted the seed that it might be fun to write a book called something like *12 Horrible Terrible Weeds and What They Want Us to Know*. I sensed I was at a turning point in my life, and I craved her advice. She was the only person in my life who shared my level of interest in journeying.

She asked to meet me at a local lake – a place I had been to only once before. (No swimming is allowed at this lake, so the beach area was deserted.) I arrived early and spent the time picking up trash, of which, sadly, there was a lot. (I decided to keep the two plastic squirt guns I found just in case I needed to defend myself one day.)

It was a seriously hot and humid day, but we found a picnic table in the shade, right next to the water. Within minutes, four black turkey vultures with red heads flew up. Three of them landed on the picnic table next to ours and stared at us curiously. One landed in the tree directly above us. They had invited themselves to our meeting!

"Death and rebirth," Lisa said, smiling. That is the traditional shamanic symbolism of Vulture. Our work together felt blessed by the birds. They stayed for a while and then flew away. After I had unloaded all my questions and concerns and issues to be "mulched," Lisa got up and walked over to what looked like a piece of trash. (I was sure I had picked up all of it earlier.)

"I thought it was this," she said, and picked up a large, beautiful brown vulture feather.

We grinned.

Nature is amazing. What's not to love? I keep that feather right next to my computer where I write, as a reminder of their blessing.

That meeting with Lisa and the vultures gave me the courage to start talking to others about this weird book idea I had. And my path became clearer.

Truth be told, though, the farther down the path of writing this book I ventured, the more I realized I had never completed my journey to Vulture. My fear had caused me to bail early. Vulture even appeared in one of my other journeys to remind me that we had unfinished business together.

I knew I had to journey again.

But before I share that journey with you, I need to explain dismemberment. It is among the many similarities found across cultures in the act of journeying. For example, a lot of the journeys I did on my own seemed bizarre until I read Sandra Ingerman's book *Shamanic Journeying: A Beginner's Guide*. The book gave me names and explanations for many things I had experienced and could not understand. One of them was dismemberment. ("Oh, *that's* what that was!" I slapped my forehead in recognition when I read about it.) Michael Pollan writes about dismemberment in his book *How to Change Your Mind* when he describes the body (and ego) experiencing death during a hallucination yet realizing that consciousness still continues on. But he calls them "ego-dissolving" experiences. It doesn't really matter what we call it, but I can assure you I have died dozens of deaths during journeys and I am still alive to tell the tale. It's only weird the first few times. The important thing to know about dismemberment is you don't feel any pain.

By now I was mostly journeying by myself, so I opened sacred space the way I usually do. I asked for protection and insight and gave thanks, all while banging a drum I "birthed" myself at a drum birthing workshop. The drum has a deep sound that changes with the weather and the seasons. I feel that striking my drum wakes up my council of elders and spirit guides and tells them it's time to come and do some fun work with me. This drumming is part of how I open sacred space, but I don't continue drumming myself during my journeys and I don't recommend it — it's too hard to focus. Journeying is an act of complete surrender.

I put down the drum and picked up the feather that Vulture (and Lisa) had gifted me. I started the drum app.

You are about to read a dismemberment journey . . .

✶ ✶ ✶

Immediately I found myself back at the same place where I had left off so many years ago. I was standing in an arid, tan, bare, sandy land. The GIANT black . . . dragon? being? . . . was still there. It had the body of a thick black snake, with pointy things running down its back like a dragon, and enormous, shiny black feathered wings, but the face of a man with a black mask that had ears like a cat. It swirled and twisted around as if it was trying to scare me. I stood my ground. This was not my first rodeo and now I knew the rules (or so I thought). He breathed fire onto me and burnt me to a crisp. Fine, I thought to myself. I'm dead. The vultures came to eat me. But then they started flying off in different directions with my bones to take back to their nests. Wait! Suddenly I felt truly dismembered. I had no form at all, even though my consciousness was still right there on the same spot where my body and bones had once been.

"You are not your body," the dragon spoke, his voice strong and authoritative. He urged me to climb onto his back, which I did. But since I no longer had a body, I looked like a transparent outline of myself. He flew upward, through two atmospheric membranes. As we passed through each membrane, his color changed to reflect the surroundings, first gray, then white.

We landed on a flat, snowy expanse. The dragon, now white, no longer had the face of a man but the face of an arctic fox (but bigger). I wasn't sure what to do next. He receded into the distance as he spoke: "The more love you feel, the bigger and brighter your energy will be. The less love you feel, the smaller and weaker your energy is."

That made sense to me, but I still wasn't sure what to do next. So I flung myself back into the snow (I did not feel cold). As I lay there, white vultures circled above me. They descended and ate my energetic body. All that was left were shiny crystals sparkling in the snow. I felt myself floating up into space and became a star in the sky. Or at least that's what it felt like. It was peaceful and quiet. Another star floated over to me and we joined our energy together. A white bird flew out from between us and down to the earth. Suddenly I was a baby in someone's arms, outside in the bright sunshine of a snowy day. I was trying to remember who I was and where I came from, but I couldn't. I couldn't remember anything. But then I looked over and saw a vulture. I remembered the vulture. The vulture was the only familiar thing.

And then the journey was over.

✳ ✳ ✳

We are not our bodies.

I mean, we are. But we aren't.

I think Vulture meant to tell us that *energy* is the truly lasting part of ourselves. And energy can take all sorts of shapes and forms. (First law of thermodynamics: Energy cannot be created or destroyed, it can only be transferred or converted from one form to another.) This journey also spoke of reincarnation, which I do tend to believe in.

But the real message was about love. The more love we feel, the bigger and brighter our energy will be. The less love we feel, the smaller and weaker our energy will be. How much time do we spend worrying about our weight, our blood pressure, our festering resentments, our jealousies, our cellulite, our status, our bank accounts, our political and personal grievances? What would happen if we spent more time focusing on feeling love?

Focusing on feeling love is, I think, the meaning of the weighing of souls, the ancient Egyptian tale of judgment upon death. In Egypt the *heart* was considered the seat of the spirit, and Ma'at, the goddess of truth, justice, and universal order, would weigh each person's heart when they died. If it was heavier than a feather, she would reject the soul, which would then be eaten by Ammit, the devourer of souls (who looks more like a dog than a vulture, by the way). If the heart was as light as a feather, that person was allowed to continue to the next world. This helped me understand something else the dragon said: "Those who are least shall be most, and those who are most shall be least." He wasn't talking about body weight.

What burdens of our hearts can we let go of?

Worrying about how we look and how much we weigh.

Finding fault with people who are different than us — whether they look different, love differently, worship differently, vote differently, or come from someplace other than where we do.

Judging others and thinking everyone else is wrong because, of course, we are always right.

Hurting others because we feel we've been hurt.

Being afraid of death.

The list is endless.

For our heart to be light as a feather, it must be filled with love alone. Then we can fly through levels and lifetimes, learning, loving, and enjoying the magic of the universe, with our heart as light as a feather. A vulture feather!

Sounds like fun to me. I'm in.

Thank you, Vulture.

Bat

Respect boundaries.

JOURNEY: JULY 29, 2021

When I was growing up, on hot summer nights my younger brother and I would run and play outside until dark, the scent of grass on our feet and our faces burnt from the sun. One of our favorite games was to throw tiny handfuls of gravel into the air and watch the bats swoop down in pursuit of the tiny stones, thinking they were bugs to eat.

I'm regretful now if I caused a bat to choke on a piece of gravel. I didn't mean to hurt any bats. They fascinated me more than they scared me. To this day, if someone says they have a bat in their house, I tell them to open a window or door, grab some flour or cornmeal and toss it out the window, and the bat will fly out after it. I tried it once and it worked.

I know there are bats living at my house because of their poop. The bats themselves are so tiny that even if I peer up where they are nestled against my dark brown porch ceiling, I can't see them. But sure enough, every year their tiny poop piles show up in the same spots — right over my front door, in one corner of my side porch, and out under the pool porch.

I only ever encountered one bat up close, face-to-face, after it fell into my pool. I stuck a stick into the water, and the bat grabbed on with its tiny claws. It was wet, so it couldn't fly, which gave me a chance to study the little creature. It bared its sharp little white teeth at me in a hiss, like a cat. Its dark black eyes and pointy nose reminded me of my dog at the time, a black Belgian shepherd (only smaller, of course). After drying off in the sun for a few hours, the bat flew off. I like to think of my bat rescue as a small reparation for all the gravel-baiting I did when I was a kid.

I am happy to host bats because they eat mosquitoes. When people visit me at my home and dusk falls and there are hardly any mosquitoes, they always ask me, "How come you don't have mosquitoes everywhere?"

"Because I have bats," I say with confidence.

I started to wonder more about them. Are my bats the same bats every year? Where do they go in winter? I looked for answers, and I discovered that bats can live for twenty to even forty years. They are the only mammal that can fly, and they breastfeed their babies (called pups). They can eat their weight in insects every night.

I'm not sure where the bats that patrol my garden go in the winter. Perhaps they hide somewhere in my house. Some bats hibernate (in what is called a hibernaculum, which sounds adorable to me), often in rock caves or old trees. Others migrate to warmer climates. Wherever my bats go, every spring when I see their tiny piles of droppings (which make good fertilizer) I am grateful they have returned. Or awakened.

Not all bats eat bugs. On my first night ever in Australia, sitting at an outdoor restaurant at a hotel in Byron Bay, I saw what looked like giant crows or hawks flying overhead, but with the unique wing shape of a bat. They were a kind of bat known as "flying foxes." Their wing spans can be up to six feet. But they don't eat mosquitoes, they eat fruit, nectar, and pollen. In fact, they are important pollinators (especially for bananas, mangoes, chocolate, and avocados — which are all *very* important things). As the Southern Cross emerged from the darkening sky and the giant bats flew from tree to tree, I really knew I wasn't in Pennsylvania anymore.

I have never understood the idea of bats as creatures to fear. There are real beings to fear in this world: rapists and pedophiles (many of whom hide behind the cloak of respectability provided by churches,

politics, and business), murderers, domestic abusers, violent crimi-
nals, cruel authoritarian leaders who start unnecessary wars. Bats are
none of these. In fact, historically speaking, there is only one thing
that kills humans more than humans kill humans, and that's mosquitoes — which, as I've said before, bats like to eat. Bats might be found in
creepy abandoned houses where terrible tragedies perhaps have taken
place in the past, but whatever happened there, it's certainly not the
bats' fault. There are at least 1,400 species of bats, and even the scariest
of them, the vampire bat, has never intentionally killed a human.

In the early days of the pandemic, bats were blamed as the source
of the COVID-19 virus. We may never know whether or not that's
true. The latest findings (at the time I wrote this book) point to
the virus starting in the Wuhan wet market in China, where many
wild animals are sold. Investigations into what exactly caused that
pandemic will continue for years. But let's not put the blame all on
bats. When nature is treated with respect, *and distance*, animals are
less likely to cause pandemics, wherever they are found. When food
is grown and handled with care and cleanliness, that food is less likely
to cause any kind of sickness.

In many shamanic traditions, bats, similarly to vultures, represent
initiation, or death and rebirth. People fear initiation and death
because it involves letting go of the old and entering into something
totally new and unknown. Perhaps the bat is the perfect symbol for
the times we live in because nature is forcing us to change the way
we live — and not gently. Flooding, pandemics, fires, heat waves,
"melting events," an atmosphere that is rapidly filling with too much
carbon dioxide and other toxic invisible gases — these are scary times.
It's normal to feel afraid. But the bat tells us is that there is life in the
darkness, food and nourishment too. And when the morning comes
and the light returns, we can rest.

Before I share my journey to visit with Bat, I need to explain something important. In the core shamanic world there are three levels:
the Lower World, the Middle World, and the Upper World.

The Lower World, or Under World, is where our animal and human spirit guides live and our subconscious roams. It's a great place to begin. Entering the Lower World starts with a sense of descending. You could jump, walk, fly, slide, or find another way in. It is not hell, although you may encounter fire, water, or challenging situations there. If you have any concerns or doubts while in the Lower World, you can ask questions of those you encounter. As Lisa says, "It is always important to ask questions of whomever/whatever you encounter. This is definitely one of my Rules of the Road. There can be trickery involved in the Under World. It is not some 'happy place' that is devoid of danger. Yes, we are in control — if we know enough to *ask questions*, keep our boundaries, and remember that we can always command whatever is frightening you to leave."

I journeyed for many years before I realized I could speak and ask questions while journeying. I wish I had known sooner, but even though I didn't, I still had amazing and enlightening experiences that gave me great insights.

The Middle World is this world we live in consciously every day, but in energetic form. You can access it by just walking out the door to your house or a place you are familiar with. However, when accessed via journeying, the Middle World is much different from the everyday world. There, humans are able to communicate with the spiritual essence of beings (like the land, rocks, trees, plants, animals).

The Upper World is where we encounter ascended masters, evolved spiritual teachers, and other enlightened beings. This world looks different to me than the Middle or Lower World; it is lighter, more ethereal. In my experience it's often filled with rainbows. To get there you can climb a tree or stairway or fly upward somehow. I've taken an elevator once or twice.

In simple terms, as Alberto Villoldo explains in *The Wisdom Wheel*: "In the lower world, we can access the ancient wisdom of the ancestors and heal the past. In the upper world, we can interact with the future, our becoming, and guides who want to help us get on a path to tomorrow better than the one we seem fated to experience. In the middle world is the realm of the present, ordinary experiences."

Each of these worlds is a complete world, including earth, sea, sky, and space. When I journey, I don't always know which world I will visit. "You should, however, set out with the intention of going to a particular world," says Lisa. "If the answer or response to your question is in a different world than you expected, Spirit will let you know." I tend to let a journey happen spontaneously. Sometimes I go down, sometimes I go up, and sometimes I just start right where I am. It's easy to navigate around these different worlds and in between them. Much easier than travel in real life. Or at least that has been my experience. This journey involves travel between worlds. Sometimes (OK, often) journeys aren't what I think they will be. This was one of them.

✳ ✳ ✳

I walked into a cave — a bat cave, if you will. Inside, a squat and scruffy cavewoman sat by a fire, beckoning me with her hand to sit down (she kept her eyes downcast). I did. She handed me some folded, cut-up leaves to eat. I did. Immediately I fell backward and shrank down to the size of a fairy. A bat appeared and told me to get on its back. I was a bit scared, but it insisted and so I did. We flew and I could see through its eyes — it was like flying through an electrical grid that bent, pushed, curved, and pulled in the darkness. It looked like a bright white spiderweb of life against a blackness as deep as space. The bat brought me back to the cave, where now a large bat was hanging upside down. It was clear to me this was my chance to speak to the bat. I think I must have been hanging upside down, too, because I had the sensation that the bat and I were both "right side up."

At first I just sat (or hung) and listened. Bat told me that we humans were invading and destroying its habitat, and that the bats just wanted to do their job and get their work done. As if there was a movie playing on the cave wall, she showed me the Congress Avenue Bridge in Austin, Texas, where a lot of bats live (it is a big tourist draw). I had seen it a few times, so I recognized it right away. She also showed me my own porch roofs and encouraged me to tell people to build more bat bridges and bat habitats.

In previous journeys I had felt more like a simple observer, but now, for the first time during a journey, I realized I could speak and ask questions. So I asked something to the effect of "What do bats do for humans?" Bat got incredibly angry.

"It's always about you, you, you. You humans are like a monster that destroys everything in your path." (As she said this, in my mind's eye I saw a Godzilla-like creature stomping through nature.) "And then you create these fantasy superheroes to entertain you and pretend you are saving the world. We are not superheroes. We are mothers. We have children. You come into our caves for entertainment and make us sick. Fuck you! We just want to be left alone to do our work. Leave us alone."

But Bat was not done.

"We eat your blood because we eat mosquitoes that suck your blood, and your blood is *garbage*. It's filled with toxins and opioids. It's made us sick so now we are making you sick. Leave us alone. We just want to do our work."

"What is our work as humans?" I asked her.

"I can't answer that for you, but maybe it's to learn to live in harmony with nature," Bat replied. "I'm tired now, so leave me alone." The sun was rising, and all the other bats returned.

Suddenly I found myself back by the fire again with the cavewoman. She got up, stabbed me, took out my heart, and ate it. Then she burped and fell asleep. The vultures came to eat what was left of me. I watched without feeling afraid, knowing this was another dismemberment. Sigh. I was then pulled up into the Upper World. The scene was filled with rainbows, and I could sense some sort of angelic spirits around me, although I couldn't see their faces or discern who they were. They told me to find the courage to speak for nature. They kissed me on the forehead and sent me back to my body.

I woke up.

* * *

Death.

Rebirth.

Speaking for nature.

These are recurring themes from my shamanic journeys. The messages were surprising to me. But the more I journey, the more I realize that nature beings have a much different perspective than I as a human do. And I am humbled by their honesty and trust in me to communicate their message.

We are a selfish species, like toddlers who leave a trail of destruction wherever we go and expect everyone else to meet our needs without our asking. Even me! I realized after the journey that I had never asked Bat what *bats'* purpose was, just assuming it was to eat bugs. And it's true that human blood, just like Earth's waters, soil, and air, has become polluted with toxins, waste, and the residues of our attempts to distract, numb, and entertain ourselves. Maybe our toxic blood has weakened bats, leaving them more vulnerable to the white-nose syndrome that has been killing them in many parts of North America. We always look to blame something else or someone else. What if it's us that is to blame?

Bats. We need them more than they need us. Please leave them alone. Respect boundaries. But also know that there truly are no boundaries between all of us. That's why we have to learn to live in harmony with nature.

Thank you, Bat.

Rabbit

Grab pleasure while you can.

I watched a jaunty juvenile fox prancing away from my yard with a limp, dead bunny hanging out of its mouth. Good job, I thought to myself. That's how it should be.

But rabbits are so cuuuuute! Baby bunnies are adorable and the sweetest thing. Until, that is, they start eating your peas, your beans, your lettuces, your bulbs, your favorite flowers. And then bunnies make you feel things that are dangerous and dark. Perhaps even delicious.

Rabbits don't just eat plants. They also eat the plastic deer fence surrounding my yard that is meant to keep deer out and chickens in. So now I lose chickens to the foxes, because chickens aren't as adept at escaping foxes as rabbits are.

Rabbits are a menace.

In Australia, rabbits are a *real* menace. White colonizers imported European rabbits for food and sport in the 1800s, leading to one of the most destructive introductions of a nonnative species in history. As you may know, rabbits are very good at reproducing. A female rabbit can bear several litters per year, with one to fourteen babies per litter. That's as many as sixty "kittens" or "kits" per mama rabbit per year. By 1920, there were more than *ten billion rabbits* in Australia, eating crops, grasses, and tree seeds and destroying habitat for the native bilbies, wombats, and wallabies (all very cute). The

Australian government built a "rabbit-proof fence" all the way across the country to keep rabbits from destroying absolutely everything. A rabbit-proof fence has to be made of hard metal, because if it's not, the rabbits will eat right through it and continue their path of voracious destruction. (In the movie *Rabbit-Proof Fence*, based on real-life events, three young Aboriginal girls who were taken from their mothers and forced into a white-led conversion school escape and follow the rabbit-proof fence all the way home. It's *definitely* worth watching.) Australia has even introduced rabbits infected with viruses and other disease organisms to control populations. As with most efforts at unnatural controls, though, that campaign has proved mostly a failure. The rabbits have developed resistance. Natural controls — that is, predators, mainly foxes and cats — have also been introduced, which makes me wonder, do three wrongs make a right?

My cat, in her prime, loved to eat baby bunnies. At first, when my family and I would hear the pathetic baby honking like a squeaky toy, we would run around and try to save the poor thing and cry over the cute little dead body. But the more I thought about it, the more I realized my cat was fulfilling her purpose.

Yes, I let my cat out of the house. Pumpkin Shirley would not be stopped. Ever. We are told not to let cats out of the house because they will kill birds. Here is what kills birds at my house: windows. Pumpkin kills bunnies and, her favorite of all, chipmunks (also cute). As Pumpkin headed into her eighteenth year I looked back and thought of all the empty cat food bags and cans and all the used kitty litter my family has added to the waste stream because of her. It made me question whether anyone should ever have a pet cat indoors. Except cats do serve a predatory purpose, whether their prey is mice, rats, rabbits, chipmunks, or birds. Everything eats. And ultimately everything wants to fulfill its purpose. And Pumpkin was very, *very* cute.

We gardeners know that managing intruders is one of our key jobs, but that does not make it easy. Fences work until critters chew through them. Raised beds will be my next endeavor to foil them. You would think I would know better by now, but I planted green beans in my ground-level beds, and the rabbits mowed them down to the stalk. How could I forget from one season to the next that rabbits will always eat my beans and peas? Fortunately I had an incredible pea crop because I planted them in — you guessed it — a *tall* raised bed.

Raw shelling peas straight off the vine are one of the greatest foods ever invented and my absolute favorite. I have to grow them because I can rarely find them in any local grocery stores. In my early gardening days, after years of failing against the street-smart small-town rabbits of Emmaus, Pennsylvania, I designed a pea protector — a movable fence with built-in trellises (because pea vines need to climb). It cost more to have that pea protector built than all the peas I could ever grow, but it was worth it because it protected my favorite snack from those rabbits. It didn't survive the move from my town garden to my hilltop garden, and was too ridiculously elaborate to construct another one.

Yes, sometimes gardening is exhausting.

As if to mock me further, while writing the manuscript for this chapter, one evening I went to "put my chickens to bed" in their coop after their day foraging in my fenced orchard. When I reached the coop, I noticed three hens were outside the fence. I brought them back in (my chickens are very tame and let me pick them up). But when I had gathered all the chickens together, I noticed one was missing. I went outside the fence again to investigate, and I saw a rabbit jump through a big hole it had just chewed through the plastic fence. (Which we had been patching and patching — this hole was a foot off the ground!) The rabbit ran off down a path, stopping only for a moment to look back at me, as if to taunt me. I searched and searched for the last chicken but found no sign of her. Poor thing.

Sometimes it feels like an endless struggle, and I decided to journey to seek some insight into the rabbit's point of view. Two days passed before I had time to do the journey, and I could sense a big brown rabbit in a dark cave. Waiting for me. Ready to talk. It was like a ghost I could see out of the corner of my eye but if I turned my head nothing was there.

* * *

"I've been waiting for you," she said.

"I know," I replied. I asked her what she wanted me to know. A plethora of baby rabbits erupted from her in a beautiful pattern, like a vision from a kaleidoscope. She got right down to business.

"We reproduce so much because we live in fear," Rabbit said. "We exist to eat and be eaten, that's why we grab as much pleasure as we can while we are alive. Your peas and green beans are like candy to us. If you don't want us to eat them, the only thing that will stop us is a metal fence, because we love eating plastic fences. It's like dental floss to us."

At that point I felt my face turn into a rabbit's face, frantically chewing and squinching my nose. This feeling lasted about a minute. It was weird.

"Also, we love sex," Rabbit continued. "Look up rabbit sex. Like I said, we grab what pleasure we can. We are not so afraid of humans because we know we have faked you out with our cuteness and soft fur. We fear foxes, cats, and coyotes. But we don't fear extinction because we reproduce so well. We love sex!"

The journey ended.

* * *

After writing down everything I remembered and having another cup of coffee, I looked up "rabbit sex." Be prepared if you put that term into a search engine. Lots of links to the "rabbit" sex toy will come up. I did find some legitimate information, but it was still on a sex site. There I learned about *superfetation*. That's when an animal (or sometimes even a human) becomes pregnant with a new batch of babies even while pregnant with another batch. This unique skill of rabbits is what may have led Christians to think that rabbits had "virgin births" and so associated them with purity.

Nothing could be further from the truth. Female rabbits, known as does, become sexually mature as early as four months old. And in an oddly hysterical way, when a male rabbit — called a buck — mounts a female, there is about thirty seconds of intense thrusting, after which his muscles relax so deeply that he literally falls over and faints.

This led me to question whether female rabbits experienced any pleasure in this process. I did come across an answer in an article on the website of *The Guardian*.* I quote:

> *To explore the question [of whether female rabbits can orgasm] the team gave 12 female rabbits a two-week course of fluoxetine (trade name Prozac) — an antidepressant known to reduce the capacity for women to orgasm — and looked at the number of eggs released after the animals had sex with a male rabbit called Frank.*
>
> *The results, published in the* Proceedings of the National Academy of Sciences, *showed that rabbits given the anti-depressants released 30 percent fewer eggs than nine rabbits that were not given Prozac but still mated with Frank.*

Frank?! Frank?! I'm sorry, I find this very funny. That Frank was one lucky rabbit. Frank, that's a lot of fucking and fainting! The goal of the study was to understand whether the female orgasm has any evolutionary purpose, which the researchers believed was related to ovulating.

> *The team said the results fit their theory that rabbits needed to experience something akin to an orgasm to have a hormonal surge and ovulate, although it is not known if it gives the animals sexual pleasure.*
>
> *They also said their theory was supported by a previous finding that animals that rely on sex-induced hormonal surges for ovulation tended to have a clitoris — the organ behind the*

---

* Nicola Davis's article "Rabbits May Hold Key to Solving Mystery of Human Female Orgasm" (September 30, 2019) is one of the (unintentionally) funniest descriptions of a science experiment I've ever read (and I've read a lot).

female orgasm — in a position that meant it was more likely to be stimulated during sex.

Well, that's interesting. Other animals besides humans have clitorises? I did not know that. I did not learn that in school. (After doing more research I discovered that *all* female mammals have clitorises and some are quite large.*) Which leads me to digress a bit about the purpose and role of journeying *for me*. Before my journey, I was feeling a little bitter about and frustrated by the plastic-nibbling rabbits in my garden. ANNOYED! I was just plain mad at rabbits. But talking with Rabbit during my journey gave me a much fuller idea of who they are and why they exist. It soothed my anger and annoyance. It made me laugh! And it led me down a "rabbit hole" on sexual pleasure that I would never have discovered otherwise.

Frank will live forever — in infamy!

The moral of this story is that I will always plant my peas and beans either in a raised bed or behind a metal fence (or both) and not waste time doing work to no avail and being disappointed and blaming the rabbits. After all, we need to grab our pleasures where we can. Just like Frank.

(Reader, you may be interested to learn I now have some three-foot-tall raised beds in my garden, which makes it easier to tend vegetable crops as I get older anyway. I also have a four-foot-tall cold frame, which provides fresh lettuce, spinach, and parsley all the way through to December. Take that, you cute little bunnies.)

Thank you, Rabbit.

---

* The largest clitoris is on the spotted hyena. It's eight inches long! Although the blue whale clitoris has yet to be measured, and it's probably larger.

Lanternfly

The way to heaven is through joy.

JOURNEY: AUGUST 14, 2021

Picture this: I filled a bucket with hot water, vinegar, soap, and cooking oil and carefully walked out to a young tree of heaven, which was the brand-spanking-new home of spotted lantern flies, an "invasive invader" from Asia. The sapling was about fifteen feet tall, weedy, with leaves that looked like ferns. My goal was to cut down the tree (with a handsaw, mind you), but first I wanted to get the giant insects off the tree and kill as many of them as possible in one fell swoop. Because we had been told over and over: If you see them, kill them. They are invasive. They are destructive and have no known predators in North America.

The laternflies were lined up together facing upward on the trunk of the tree. From a distance they look like bark — a part of the tree — but closer inspection reveals them as a swarm of bugs. When their wings are closed, they are the same color as the tree trunk. But when they fly you see their red-and-white wings in a whirl of color. Each insect is about an inch long.

I flung the water at the tree and suddenly about four hundred lantern flies erupted in every direction, including straight toward my face. They don't bite, but they are *big*. I SCREAMED! (My kids will tell you I'm a screamer.) I ran as fast and far as I could. I felt myself starting to fall right as I reached my macadam driveway, which is truly not an ideal place to land. Somehow, I managed to stumble all the way across the driveway without falling until I got to the grass on the other side. There I tumbled down inelegantly, still holding the bucket, laughing so hard, only slightly injured, and triumphant that there was still an inch or two of my nontoxic death fluid in the bucket. It was my version of a slow-motion English Premier League football fall (officially known as a "dive"). I was very proud of it.

Immediately Lucia, who witnessed the whole thing, did a reenactment, and we laughed some more while the dog, Penny, panted happily, just excited to be in the front yard.

Returning to the tree, I started stomping on the still quivering bugs, determined to obliterate all of them. After I was satisfied that most of them were dead, I sat on a rock and started to saw through the trunk. The tree of heaven is a weedy tree — the kind that grows through cracks in the pavement in cities. It's a prolific producer of new sprouts from its seeds. I had cut down a large tree of heaven about ten years ago, and it is still one of the most common "weeds" I pull from my garden. The sapling I was sawing was about six inches in diameter — a "weed" that got away. Sawing with a handsaw takes patience and, since I'm getting older, a break or two to rest. A third of the way through, I thought about calling someone with a chain saw to help. But I persisted. Two-thirds of the way through, I took another break and sat on the grass and pondered this spotted lanternfly.

It probably didn't choose to come to this continent. Maybe it got packed into a crate of stuff that someone was sending by accident. It could be a new military tactic sent by the government of China to destroy the United States, but probably not. More likely it snuck into a case of cheap merchandise headed to a Dollar Store. If it wasn't labeled as a terribly invasive species, it might even be beautiful. Even though its eyes are bright red and scary.

But it was while stomping a few of them to death that I thought of those videos you see on social media — you know the ones — of a man, maybe a cop or a white supremacist, stomping and kicking a person who is on the ground huddled in a ball and crying for mercy.

Wait. Is that what I was doing? How was I different? I was aggressively killing something "that doesn't belong here" even though it was no true harm to me. Because I was told "If you see them, kill them. They are invasive!"

We went through this a decade ago with the brown marmorated stink bug, another insect accidentally imported to North America from Asia and first spotted in the part of Pennsylvania where I live. But that was slightly different because those stink bugs would show up inside the house. I remember one warm day in late September when I vacuumed up thousands of them *from my bedroom*. Not cool. That's breaking and entering, if you ask me. (But we are friends now, Stinky and me.)

The spotted lanternflies weren't in my bedroom, though. They were in my front yard, which, let's face it, is really the woods. There is no controlling The Woods.

I thought about what the Indigenous Americans must have felt, seeing foreign white strangers with weirdly colored eyes swarming their land, stealing their food, and killing their families. Those same European colonizers brought a strange new disease (smallpox) that ended up killing an estimated *90 percent* of all Indigenous Americans. What the colonizers did is beyond tragic. Fortunately, a vaccine for smallpox was developed and the disease was finally eradicated in 1979. But that was a few hundred years too late to save the Indigenous people of Turtle Island, the indigenous name for what colonizers called America.

I finished cutting down the tree and walked back into my house, feeling slightly ashamed of myself. But I also appreciated the insight into my own behavior and the way in which it is so easy to get caught up in frenzies and swarms of actions that don't make much sense. I consider myself a peaceful, nature-loving human, but even I could turn vicious given the right circumstances. How and why does that happen? I'm not sure. Ultimately most of us are immigrants. And many of us and our ancestors were vilified on arrival. Some arrived voluntarily, others involuntarily. But no matter when or who, the story is the same.

Within a few years of lanternflies' initial wave of immigration to the United States — and of my attempt at their massacre — they seem to have settled into the environment, just like the stink bugs. I rarely see them anymore. That could be because people expended so much time, energy, and toxic spray trying to kill them — I hate to think of the pesticide residues left behind and what effects they are having on our bodies. Or perhaps the insects have assimilated and now they are one of us, part of that old American melting pot, which we once celebrated as our special gift to the world, as that characteristic that makes our culture so unique.

I watch the culture wars of today with a sense of confusion and sadness. People on all sides seem to throw whatever horrible words they can at each other without showing any desire to understand, listen, or resolve differences. It's exhausting and mean.

With that sorrow weighing me down, and the lingering memory of my own attack on the lanternflies in the tree of heaven, I felt compelled to journey to ask Lanternfly what it all means, even though I'm a little creeped out at the idea of journeying to talk to insects.

I don't like to journey when there are other people in my house, and my kids were staying with me for two weeks of summer fun. Halfway through those two weeks my beloved but elderly cat Pumpkin died. I decided to wait until everyone left before I would try to visit with Lanternfly. I found myself hoping that Pumpkin would show up in my journey too.

On an early Saturday morning, in my now-empty house, I smudged with sage from my own garden, opened sacred space, and lay down on my magic blanket.

The drumming started.

✳ ✳ ✳

I opened a wooden door in a tree and entered a meadow and woods. I smelled incense and felt that I was somewhere in Asia (in the distance I could see a temple). I was very small and hidden among the grasses. I had become a lanternfly. A little Asian girl leaned closer to play with me. I hopped, and she laughed and hopped after me. (During one nymph stage of a lanternfly's development, before it grows wings, it is a bright red jumping bug with white and black spots. But even when they are mature, lanternflies hop as much as they fly.)

Lanternfly spoke: "In Asia we are symbols of joy. Children love to play with us since children do not judge. They have to be taught to judge." I saw the small child hopping after a lanternfly nymph and giggling.

"We came to America as eggs on wood. Apparently Americans need more joy. But we were met with anger, fear, and hatred." I heard the word "killjoys" whispered around me, as if the lanternflies were accusing Americans under their breath of being destroyers of pleasure and no fun at all.

"We line up on trees of heaven to tell you that the way to heaven is through joy!"

At this point in the journey, I started sobbing. Of course. How had I not known this about lanternflies before? Even calling the tree of heaven a "junk tree," as I had, seemed horribly wrong.

"Our purpose is to bring joy. In Asia we are considered beautiful."

Then Pumpkin showed up, nudging me and licking the tip of my nose like she used to. She proudly showed me her baby angel wings. She said, "Pets exist to teach you humans how to love — but you need to learn to share that love with other humans. We all have a purpose, and joy exists when everyone and everything is living fully in their own unique purpose."

✶　✶　✶

I came out of the journey stunned and humbled. In retrospect these things seem so simple and obvious, and yet we are so blind to them. (*I* was so blind to them.) It takes a journey for me to see things as they really are, not just as I imagine them to be. The tree of heaven, also known as the Chinese sumac, the stinking sumac, or stink tree, was brought to the United States from China in the late 1700s as a shade tree. It was popular in cities because it was so easy to grow. In fact, it's the tree that is featured in the novel *A Tree Grows in Brooklyn* by Betty Smith:

> *She looked down into the yard. The tree whose leaf umbrellas had curled around, under and over her fire escape had been cut down because the housewives complained that wash on the lines got entangled in its branches. The landlord had sent two men and they had chopped it down.*
>
> *But the tree hadn't died . . . it hadn't died.*
>
> *A new tree had grown from the stump and its trunk had grown along the ground until it reached a place where there were no wash lines above it. Then it had started to grow towards the sky again.*
>
> *Annie, the fir tree, that the Nolans had cherished with waterings and manurings, had long since sickened and died. But this tree in the yard — the tree that men chopped down . . . this tree that they built a bonfire around, trying to burn up its stump — this tree lived!*
>
> *It lived! And nothing could destroy it.*

I went outside to Pumpkin's grave to thank her for showing up in my journey. There is a beautiful rock to sit on there, placed by my son-in-law, an immigrant. And so I sat, just absorbing this new insight into life and feeling very grateful. After a few minutes I looked down at the ground. There, on a small blade of grass, pointing toward heaven, was a single lanternfly.

Thank you, Lanternfly.

Lightning Bug

*Find the beauty within you
and let it shine.*

JOURNEY: AUGUST 22, 2021

Lightning bugs are not pests, and my original plan was to write a book about plants, animals, birds, and insects that I, or others, find pesky and difficult to love. But once I started journeying to listen to the spirits of annoying beings like rabbits and lanternflies, I began to realize that there were crowds of others who were most anxious to connect — I felt that they were begging me to talk to them. I even started keeping a waiting list. Lightning Bug was especially persistent.

Remember the day I sat by the lake with Lisa and the vultures showed up? While Lisa and I were talking, three or four lightning bugs landed on me (in broad daylight). Then I heard a loud clicking in my ear. I asked Lisa to see what was tangled in my gray frizzy hair, and it was, you guessed it, another lightning bug. It had been *shouting* in my ear with its clicks. Since the vultures were a literally larger presence that day, I made a mental note to check in with the lightning bugs later.

A few months later, while I was at home one evening, a lightning bug aggressively got in my face again, as if to say, "Here I am! Talk to meeeeee!" During summer I often leave the sliding doors and screen doors of my house wide open so people and animals can come and go at will. (I don't have air conditioning.) As darkness fell, we were watching the Olympics on TV. A lightning bug showed up and flitted about my kitchen in front of the TV. *This happened three nights in a row.* Perhaps it was just drawn to the bright light of the TV, but it's not unusual for us to turn on the TV in the evening, and this had never happened before. "OK, OK," I said. "I will visit with you and see what you have to say."

I wondered if the lightning bug's presence in my kitchen had to do with my granddaughter. She was visiting at the time and, being four years old, was in a prime lightning-bug-catching phase. Just that day she had said, "When I was a baby, I thought that plants were nature. Now I know that everything is nature."

"Did you learn that at camp?" I asked, amazed that she would learn it at school.

"No, I learned it in your garden," she said matter-of-factly.

That, my friends, is why I have my magic garden.

On a rainy Sunday morning of a full moon in August, when I didn't have any visitors, I prepared for another journey. I couldn't decide whether to set an intention to talk to lightning bugs or to yews, which had also been prodding me to speak to them. I invited them both to tell me what they wanted me to know. I smudged with sage from last year's garden. I lit some candles. I drummed on my drum while I opened sacred space. Then I lay down and turned on the drumming app . . .

✳ ✳ ✳

It was a blue dusk and foggy. I walked down a path through a cemetery. At the end of the path was a giant wrought-iron gate with two large yews on either side. I walked through the gate. Suddenly I was surrounded by lightning bugs and their clicking. I even started to make clicking noises with my own mouth. I had become a lightning bug, and I flew around, landing on the branch of a yew. Then I also felt myself rising up in the air and then landing on the ground, over and over, up, down, up down. The grass and soil were where I went to rest. The sky and trees were where I went to shine. I could smell the soil and feel the trees.

Then I started to hear the soft voice of Lightning Bug: "We exist to inspire magic and show that true beauty isn't how you look, but the light you give off. We are beautiful. You are beautiful, and we love you, but can you please tell people to stop using poisons? It's killing the magic and killing us. What is the price of love? Is it worth it? Just to make some guy rich?"

Lightning Bug showed me images of people who were wealthy and superficially beautiful, but I felt that many were empty inside. There were people who made money by selling chemicals that promise to create the illusion of beauty — lawn and garden chemicals, farm chemicals, cosmetic chemicals, forever chemicals like PFAS. These people did not give off light. They radiated a hunger for money, status, and attention that could never be satiated.

"Find the beauty within you and let it shine," Lightning Bug said sweetly and softly.

Then the lightning bugs led me to a boat and we crossed a river in the dark. It was a beautiful procession of lights, reflected on the dark water. Yes, I knew it was the river of death and that Yew, a giant presence, was waiting to greet me on the other side, but I wasn't afraid. Lightning Bug was reassuring me that everything was beautiful and fine. I was curious to hear what Yew would have to say. But that's a story for the next chapter . . .

* * *

Lightning Bug's message resonated with me. The truth is I have often felt insecure about how I look. And judged. And judging of others even when I know that's a terrible attitude to have. When I was climbing the corporate ladder, I wore makeup and fancy clothes, even though it felt false. If I appeared on TV or had "headshots" taken, the corporate communications teams would bring in stylists to make me "look the part" even though it made me feel gross. All too often, when I met someone for the first time they would comment, "You don't look anything like your photo!" I realized I needed to decide who I was going to be — the person in the publicity photos or the real me. I thought about all the male executives who never wore makeup or got their hair highlighted and straightened (although quite a few were dyeing their hair). I calculated all the time, money, and kilowatts I would save by not dyeing and blow-drying my hair or putting on makeup every day, and I decided it wasn't worth it. I stopped doing it. I stopped painting my face. I stopped trying to control my hair. I stopped wearing shoes that hurt my feet and knees.

And then I started to notice something else. Many of the beautiful, rich, and seemingly perfect-looking people I got to know were often unhappy, insecure, and suffering from eating disorders, anxiety, or depression. Physical beauty was not the ticket to happiness (or health) I had assumed it was. I would often walk down New York City streets in the company of intensely beautiful people and see how other people stared at, drooled over, and worshipped them. (As for me, I kind of like being invisible.) Those memories make me glad to no longer be part of the media business that feeds off people's insecurities, creating a veneer of beauty that conceals a sense of emptiness that is longing to be filled.

Beauty is important. Beauty is primal. The universe overflows with beauty. Perhaps that's why so many people aspire to be "beautiful" through cosmetics, clothing, body shape, and adornment. But being rich and beautiful does not protect you from heartbreak or suffering. Ultimately true beauty comes from within, through letting your light shine brightly.

I find that the more I focus on my own authentic unique gifts, as quirky as they are, the more I feel alive and able to shine my light for others . . . without artificial chemicals, without makeup, and without having a thin and toned body (a little fat is actually good for you). And I've found that it's my inner light that attracts the people who become my true friends.

Now, about the light that lightning bugs create . . . Lightning bugs, sometimes called fireflies, are a group of insect species that light up at night by a process called bioluminescence — ingested air interacts with a chemical called luciferin that lightning bugs produce in their bodies to make light. (Fun fact: the name "Lucifer" derives from Latin and means "light bringer." Somehow, somewhere, that turned into something bad.) Their light is their language of love to attract mates — or sometimes other species of firefly to eat.

Despite their name, fireflies are not flies but beetles. They are slow fliers, which is why they are so delightfully easy to catch. And there are two thousand different species. They spend the majority of their life (sometimes a few years) as larvae underground or under tree bark, where they feed on slugs and other bugs. They love rotted logs, wild grasses, clean (not poisoned soil), natural sources of water, and darkness. (Sounds a bit like an "in search of" ad, doesn't it?)

And yes, firefly populations are declining — drastically. In part that's due to all of the chemicals we use on farm fields, on lawns and sidewalks, on gardens, on killing all those "invasives." But it's also due to light pollution. So . . . stop giving your hard-earned cash to chemical companies that kill the magic and beauty of nature and the magic and beauty of *you*. Stop worrying so much about detoxing yourself and work on detoxing Earth. Turn out your lights, go outside, and enjoy nature's light show. Feel your own light grow and start to glow.

Thank you, Lightning Bug.

Yew

Let yourself be wild.

JOURNEY: AUGUST 22, 2021

When I was very young, perhaps three years old, I held a red, gelatinous yew berry in my hand. My older cousin shouted, "Don't eat that berry!" So, of course, I popped it right into my mouth.

I can still see the stainless-steel bowl in the pale green hospital room where I was told I needed to throw up. They gave me syrup of ipecac. I was sitting on someone's lap. I don't remember throwing up. But I survived. And I will never forget that experience.

Yes, yew berries are poisonous. Actually, only the seeds inside the berry are poisonous. Still, don't eat the berries unless you really know what you are doing. (Either way, I wouldn't if I were you.)

After my cat Pumpkin died, I wanted to plant a bush by her grave, and for some reason I kept thinking about yews. Yews would not stop popping into my head.

Where I live in Pennsylvania, yews have a common and, to me, sometimes comical role in the landscape. Almost every traditional-style house, no matter how big or how small, has at least one yew in the front yard, pruned to its barest shape. Sometimes it's pruned as a box. Sometimes it's pruned as a ball. Sometimes it's pruned crooked, but hardly ever is it not pruned at all.

The first time I ever noticed unpruned yews was at our local Wildlands Conservancy preserve, where there is a circular grove of unpruned yews that must be almost a hundred years old. It creates a magical shelter, and inside are benches to sit on and read stories to kids. It is one of the most magical spaces I've ever seen. Of course I had to plant a circle of yews in my own yard, and I never prune them. They surround my trampoline, so in the short term their purpose is to hide the trampoline from view, but longer term, after the trampoline is gone, it will be my own magical circle of yews.

My theory is that the obsession with pruning is a throwback to the idea that a person's home is their castle, and that humans can control nature. It reminds me of the incredible castle and garden Château de Villandry in the Loire Valley in France. When I took my kids there to see it, they loved the intricate mazes of hedges, but those mazes reflect the huge amount of money that the château owner expended to control nature through pruning in an effort to impress the neighbors and entertain guests. It's beautiful, fun, and gorgeous. But as a gardener myself, I understand that this amount of work cannot be done without quite a large staff. Sadly, it was probably slaves or servants who originally maintained it.

In descriptions of yews, you find two common themes: (a) They are toxic, and (b) They were often planted in graveyards to keep out livestock. Ahh, that makes sense. I will get a yew to plant by Pumpkin's grave.

At my local nursery, I was drawn to a plant called Japanese plum yew, specifically a cultivar called Duke Gardens. Since my oldest daughter is a historical romance writer, I'm always drawn to anything that has a duke involved. But when I got home and looked up the Latin name — *Cephalotaxus harringtonia* — of my new shrub, I realized I'd been faked out. It's not a "true" yew; whose genus is *Taxus*. But I planted the plum yew by Pumpkin's grave anyway.

The night before my planned journey to visit Lightning Bug or Yew, I was reading *Finding the Mother Tree*, the poignant and fascinating memoir by forest ecologist Suzanne Simard, which illustrated the frustrating resistance and slowness to change in the science and forestry community. I came across this passage in which she writes about her breast cancer treatment:

> *Dr. Malpass was right. The paclitaxel infusions were easier to absorb than the earlier chemo drugs, and I regained some energy and started to walk in the forest again. Paclitaxel is derived from the cambium of the yew — a short, shrubby tree that grows under old cedars and maples and firs. The Aboriginal people knew its potency, making infusions and poultices to treat illness, rubbing its seeds on their skin for strength, bathing in preparations to cleanse their bodies. They used this tree to make bowls and combs and snowshoes, and to craft hooks and spears and arrows. When the anticancer qualities of the yew were brought to the attention of the modern pharmaceutical industry, there was a bounty on the trees. I'd find the small yews — their branches as long as their stems — stripped naked of their bark, looking like crosses, specters of maltreatment. In recent years, pharmaceutical labs have learned to synthesize paclitaxel artificially, leaving the yews to thrive under the cool canopy of the forests. When the old growth is clear-cut for the big timbers, however, these small scaly trees are left weakened in the hot sun.*

Magic, I tell you. Here is part two of my journey to Lightning Bug and Yew.

✶ ✶ ✶

The lightning bugs ceremoniously escorted me across the river of death to a giant Yew on the other side. She was enormous and sinewy — the mother of all yews. I went into her arms, and she embraced me. It was a little tight, but also comforting.

She said, "When you die, my roots will surround you and transform you. Death will heal you. Death is the ultimate healer."

At this point, as I lay on my magic blanket, my arms reached up and wove about like her branches, as if I were Yew. I spoke aloud:

"When you let yourself be wild, only then can you truly live." I put my hands on my heart and cried.

Yew continued, "People are always trying to control us [yews], to keep us in boxes. But it's them that are boxed in. They are building their own coffins." She sighed. "Humans are so exhausting."

I lay down underneath Yew and watched the fireflies lighting up the night. It got very quiet. Then, beneath Yew, Pumpkin appeared. She came over to me with her little wings and sat on my heart, purring.

"It's all about love, Maria. It's all about love. And everything is beautiful," Pumpkin said softly to me.

I cried some more.

"And also," said Yew, "plant two *real* yews next to Pumpkin's grave. One on either side."

✳ ✳ ✳

I will. You know I will.

And I did.

But still, I can't help wondering about all the pruning people do. Why are people so afraid of wildness? Of life?

Through my work in publishing and media, I learned a lot about people — by being a manager and by studying customer research. I also learned a lot simply by talking with neighbors and family. One thing I noticed is that we are all afraid. We are afraid of what we don't know. We are afraid of the future. We are afraid of change. We are afraid of bugs, germs, and diseases. We are afraid of what our neighbors will think. We are afraid of going to hell. We are afraid of looking weird or fat or of being different. We are afraid of not having enough money or friends — or if we have enough, we are afraid that someone will steal what we have. We are afraid of being alone and going places by ourselves. We are afraid of not being enough (although some of us are afraid of being too much). We are afraid of being teased, made fun of, or worse, rejected. We are afraid of speaking up. We are afraid of feeling vulnerable and ashamed. We are afraid of not being loved. We are afraid of death. We are afraid of dying.

Our fears often play out in our landscapes, whether it's the lawn that's perfectly manicured, the bush that's meticulously pruned, the use of insecticides to kill pests that scare us, or the use of herbicides to kill weeds that embarrass us. The war on weeds. The war on bugs. The war on wildness. The metaphor of war on anything puts us into a position of offense or defense, perpetrator or victim, with a constant fear of losing. Failing. Doing the wrong thing. Losing control.

Sometimes I am still afraid too. I was afraid to talk to insects during a journey, for example, but as it turned out, the experience wasn't bad, it was *wonderful*. Facing my fears head-on — by observing and working with nature up close, getting to speak with the things that scare me so that I can understand them instead — has enabled me to embrace wildness in a whole new way. I can now appreciate the messy wildness of a young forest that is healing from trauma. I can smile at the unruly plants in my garden and understand that they are just trying to help me and do their important work. I see firsthand that nature collaborates more than competes. I see that nature is confident and patient. And that, my friends, has made me more comfortable and accepting of my own wildness. I am much more likely to laugh at myself. I see now that much of the stress I have felt in life was unnecessary and of my own invention.

I am happy to see the "rewilding" movement in environmental conservation and agriculture worlds. It aims to allow nature to do its own thing and encourages predators and other essential species to return and do their work. It takes a "passive" approach to managing wild areas, and I completely support it. I can't tell you how many people I've met over the years who believe it is our job to "manage" nature, especially forests. I'm starting to think our real job is to learn to manage *ourselves* — but by "managing," I mean freeing ourselves from so many self-imposed constraints. (Although we white people have a lot to learn from how Indigenous people were stewarding their land before we so rudely interrupted them, thinking we knew better. We didn't.)

Every single day we are alive is a gift. Perhaps death *is* the ultimate healer. But for now, I'd rather be alive. I will embrace my wildness.

Pumpkin's grave, the two yews, and the Duke Gardens bush are right outside the window by my writing desk. She used to sleep on the chair next to me when I wrote. Now she sleeps outside, buried in a basket. As of today, the bushes are tiny — about a foot tall. But I know they will grow. And I have no desire to prune them. Their wildness will be a reminder to me that when I let myself be wild, then, and only then, am I truly alive.

Thank you, Yew.

Osage Orange

Hurry. Wake up.

My first awareness of the Osage orange was when my former brother-in-law made a bow from its wood. I watched him work on it and listened to him talk about how strong the wood is. He killed deer using that bow and hung them upside down on the side porch of the house on the main street of Emmaus where we lived in the 1980s. The sight of a dead doe hanging from our porch was a bit shocking, but the venison sausage was delicious.

My second awareness was many years later, after I had moved to my house in the woods and had to drive Maya to school each day. Along the way we passed a line of huge, deeply ridged Osage orange trees by the side of the road. I knew they were Osage orange because every fall the giant lime-green balls would line the road and gradually get smooshed by cars driving by.

I wanted to stop and pick up some of the balls, but I was scared. The road didn't have a lot of traffic, but the cars moved fast and there was no good place to pull off. It took about ten years before I had the "balls" to stop (I was driving Eve to school by then). I dashed across the road, grabbed as many as I could, and put them in the trunk of my car. The entire car filled with their intense fragrance — citrusy, herbal, clean and bright.

Later that day, as I ran some errands, a young man came out to my car to put something in the trunk for me. I will never forget the look on his face as he stared at the Osage orange balls and stilled. After a minute he asked, "What are those? It's the best smell I've ever smelled in my whole life."

I tried planting those balls. But the truth is I can be totally lazy. By "planting" I mean I threw them into the weeds at the edge of my woods, hoping nature would do its thing. Nothing grew (at least that I know of). Occasionally I would stop by the roadside row of Osage oranges and grab a few more balls. I would keep them awhile to enjoy the smell and then put them on my compost pile when they started to rot.

Fast-forward a decade. My doorbell rang. I opened the door and there was a masked man I did not know. But it was pandemic times, so the mask didn't alarm me.

"Pardon me for stopping by," he said, "but I was wondering if I could walk through your woods over by the park. I've seen some really interesting trees and I want to make sure I have permission to go check them out."

"It's not my woods," I explain. "The land belongs to the local conservancy, so of course you can go there. What trees?"

He mentioned ironwood and started to explain a bit. I smiled. "Yes, I know about ironwood." He was shocked that I'd heard of ironwood trees. Before you know it, we were talking trees and I could tell he was really into them.

"Have you ever heard of Osage orange?" he asked.

I gasped. "I love Osage orange!"

"Well, you know those seeds, right?" [Indeed I do.] "I have been planting them and they are growing! I've got about forty of them. Do you want some?"

Do I ever!

He brought me six of them growing in Styrofoam coffee cups with the bottoms cut out so the roots could spread. I planted them in a special place, near where one of my dogs is buried. Then I dropped off a dozen eggs for him as thanks.

A year passed. The doorbell rang. (It was morning, and I was still in my flowing Indian caftan and my frizzy hair was wild. I'm sure I looked like a witch. I am not a witch.) It was the Osage orange guy.

"I don't know if you remember me," he said hesitantly.

"Yes, I do."

"I was wondering if you or the conservancy would be interested in the rest of my trees. My wife and I are being deployed to do some hurricane relief work, and I may not be coming back and want to make sure they have a home."

It turned out he's a Jehovah's Witness (which explains his comfort with knocking on strangers' doors), and he and his wife would be traveling down South to help with Hurricane Ida cleanup (that hurricane did massive damage in Louisiana and caused flooding in many states). I told him that as long as he promised not to try to convert me to being a Jehovah's Witness, I'd be more than happy to find homes for his "refutrees." I didn't hear from him for a few weeks, and I thought I might have offended him. Or he forgot. Or his plans changed.

I began to realize that Osage Orange was trying to get in touch with me. I decided to do a journey and find out what the message was. I had never done a journey to meet a tree spirit before, although trees featured prominently in almost all of my journeys. I was both wary and curious.

✳   ✳   ✳

I began my journey by stepping between two of the familiar Osage orange trees along the road I had driven so many times. (Lisa said this sounded like a Middle World journey. In hindsight, I agree.) As I stepped through, I realized I was in the middle of a "hedge apple" fight. (Hedge apples are a common name for the green balls.) Kids were laughing and screaming and throwing the balls at each other in a game as old as humans.

Suddenly I shapeshifted into the tree. I felt myself growing taller and towering over the kids, and my arms reached out like branches. I spoke out loud in the voice of the tree (calm, a bit weary): "I am strong. I am patient. But humans need to stop fighting. They need to find a new game to play. War is a stupid game that children play. Find a new game and don't use me for it."

"Why am I drawn to you?" I asked.

"I am beautiful and powerful," said Osage Orange. "You like me because I protect you and make you feel safe. We trees are ancient. We hold the secrets. We were here long before you and will be here long after you. We see everything. When humans learn to honor and respect trees, then they will truly learn how to live."

"How can we do that?" I asked.

"I want to be used properly. Use my dead branches to burn and my trunk to make things that bring joy — shelter and protection, not weapons."

Suddenly I started to blow huge gusts of air out of my mouth. In and out. In and out.

"Your breath feeds me, and I transform it into oxygen that feeds you."

"Why are your fruits so beautiful and why do they smell so good?"

"When you can see our fruits as beautiful rather than weapons or a nuisance is when you begin to awaken. Your friend is awakening. Don't worry, we appreciate you and will protect you. But *hurry*. Wake up."

* * *

The journey ended, and I ate breakfast and went outside to feed my chickens. It was one of those beautiful September mornings — sunny, clear, cool, and fresh. After I gathered eggs, Elvin came out to say good morning. Elvin lives in the apartment above my garage. He works for me doing projects around the house and yard and taking care of the chickens when I am away. He is Taino Puerto Rican and grew up in the South Bronx and is now retired after a career as a corrections officer in New York City. I met him through our yoga teacher, and he is a very gentle soul (who has a black belt in karate). We talked about the pandemic and the hurricane.

"I don't know why, but I feel so protected up here," Elvin said. He told me that when he goes into the nearby cities he feels constantly on guard, for both Covid and racial reasons. "But up here I feel very safe. Even all the wild animals are so happy. I can see it in their eyes."

"Thank you," I said. "This is what I do. One day the whole world could be like this. But we have to appreciate and listen to nature and each other to make it so." I've learned from Elvin that much of the safety I experience as an older white woman is not shared by the brown people among us. Even simple things like knocking on someone's door can be rife with risk. Fortunately, Elvin and I make a good team.

The name Elvin, by the way, is Irish for "magical elf." As Lisa says: YCMTSU. You can't make this stuff up.

I went inside, sat down, and started writing this chapter.

A few weeks later, my neighbor, the tree guy, stopped by again, and he told me he would be helping people in New Jersey who lost their homes to tornadoes rather than going down South. He gave me about thirty little Osage orange refutrees. I planted a dozen of them in my front yard and took the rest to the Rodale Institute, where there was plenty of land to plant them. From the institute staff, I learned that the range of the Osage orange is shrinking because the only creature that can eat the big green balls of seeds, digest them, transport them someplace new, and then poop out the seeds is . . . the woolly mammoth. Woolly mammoths used to roam the North American continent in herds. They went extinct four thousand years ago. It is debated whether they became extinct due to hunting or climate change. We may never know.

Scientists in South Korea are trying to bring the woolly mammoth back into existence through cloning. And Russians are digging up dead woolly mammoth bones as the permafrost of the Siberian tundra starts to melt. The tusks are sold to make trinkets and art that used to be made from elephant tusks, which are now illegal to sell because of concern about elephant survival. Is it right to clone an extinct animal? Who can say for sure? Is it wrong to sell woolly mammoth bones for trinkets? We may never know about that either. But humans insist on doing all sorts of things whose impact — for better or worse — can't be known until long after the fact.

The woolly mammoth is a reminder that nothing about the future is guaranteed. Not our livelihoods. Not our wealth. Not our health. Not human existence. Maybe a few thousand years from now some other being will be digging up our bones to decorate their houses and make into trinkets. We humans have been so distracted fighting against each other for the most ridiculous reasons that we have failed to see the truth right in front of us: This life, this planet, and the world we have created on it is a gift. The greatest and longest-lasting joy comes from creating good things and caring for each other — despite our differences. There is room in the world for Jehovah's Witnesses like my neighbor *and* spiritual mutts like me. It's the trees that connect and unite us.

And the trees will outlive us all.

Hurry! Wake up!

Thank you, Osage Orange.

Mosquito

When we kill others, we kill ourselves.

FIRST JOURNEY: SEPTEMBER 9, 2021
SECOND JOURNEY: OCTOBER 28, 2021

The truth is, even though I have lots of resident bats, there are still *some* mosquitoes around my yard every summer. And of course they annoy me. They come out after big rains, and at those times I am never far from a tube of anti-itch cream, especially toward the end of summer. The waning of hungry late-summer mosquitoes is one of the only reasons I look forward to fall.

I am constantly traipsing about my garden looking for pots or toys that might contain standing water where mosquitoes could lay eggs. The worst culprit is old tires. It never ceases to amaze me how many old tires I can find in the woods that surround my house and garden. Over the past two decades I've found and removed at least a hundred. And new ones keep appearing. Like mushrooms (and mosquitoes) after a rain. I recently discovered a pile of about fifteen partially covered with moss. It's my spring project to dig them out and send them off to wherever old tires are *supposed* to go, wherever that mysterious place may be. Old tires are NOT supposed to be dumped into the woods on the side of the road. People, stop doing it! Old tires are mosquito breeding factories.

Apparently, the deadliest being on planet Earth is not the wolf, the grizzly, the great white shark, or even the human (he comes in second). It's the teeny-tiny buzzing annoying mosquito. When it bites you, it first injects an anticoagulant into your body so it doesn't choke on blood clots as it sucks out your blood. The anticoagulant is what makes you itch. As it feeds, it also injects its saliva into your bloodstream, and that saliva can carry organisms that cause potentially fatal diseases. That's what makes it so dangerous: mosquito spit.

I confess, I was nervous to visit with mosquitoes in a journey, but they were on my list to talk to because they are so annoying (and deadly) to so many people. I can't say I was looking forward to the encounter, but I wanted to ask mosquitoes why they exist. It was hard for me to imagine a good reason why, and yet I was pretty sure there must be a good reason. So I just dove in . . . literally.

✳ ✳ ✳

It's not unusual for me to have a mental image of walking or jumping into water at the beginning of a journey. Basically, I go wherever my mind's eye leads me. As I entered this journey, I was in a tropical jungle, and a dark murky pool lay before me. Believe me, I did not want to go in there. But rationally, I knew I would be fine — I wouldn't wake up from my journey wet and mucky. After all, I was safe on my sofa. This pool of water is what had appeared before me as the place to go, and in I went.

As soon as I dove in, I transformed into a squirmy mosquito larva. Then I floated to the surface, emerged from the water, and flew. Somehow, I homed in on the back of a sweaty human male and drank his blood. It was delicious. Warm. I mated. Laid eggs. And died.

I was dead, and nothing was happening.

Ummm . . . hello? Hello?! Nothing seemed to be happening. Into the void I called out, "Take me to your leader!" Weirdly, I found myself transported to what I can only describe as a spaceship command center, and a giant mosquito was before me.

"We've been at war with humans ever since you arrived," it said.

"But don't you need us to eat?" I asked.

"We tend to kill what we love."

"Huh?"

"We tend to kill what we love. There are too many people anyway. Do your homework and come back to me. But why should I help you? So you can eradicate us?"

This was turning into a really strange journey, and I felt unwelcome. As I turned to go, Mosquito stuck its proboscis into my head and injected something into my brain (probably spit). It repeated: "Come back after you've done your homework. We tend to kill what we love."

* * *

I came out of the journey, and I had to admit that it was a fair criticism. I had not done my homework; I really knew almost nothing about mosquitoes other than that they were quite annoying and potentially deadly. I thought about the statement "we tend to kill what we love" and found myself wondering: Aren't many human murders due to domestic violence — fits of rage, jealousy, honor killings? We tend to kill what we love. But how did that relate to mosquitoes?

I started reading, and one of the first things I learned is that mosquitoes are old. Like at least 236 million years old. That's a lot older than humans.

Only the infected female *Anopheles* mosquito causes malaria. In fact, among virtually all mosquitoes (and there are more than three thousand species), only the females bite. And it's not just malaria they spread. It's West Nile fever, yellow fever, chikungunya, Zika fever, and dengue. These diseases are responsible for about 5 percent of annual deaths of humans worldwide — almost a million a year. (Although antibiotic-resistant diseases are rapidly gaining in power to kill due to the misuse of antibiotics to fatten up livestock.)

In truth, the human death rate due to murder is close to the death rate due to malaria. And as I write this manuscript, the annual death rate due to COVID-19 is way higher than deaths due to mosquito-borne diseases and murder combined. We live in a deadly world. And mosquitoes are just one tiny reminder of it.

In the warm areas of the northern hemisphere, mosquitoes have been primarily a nuisance, not a threat to human life. But as global travel and trade have increased, average temperatures continue to rise, and humans create even more habitats where mosquitoes thrive, tropical diseases are spreading rapidly.

All this is interesting, but I kept returning to this: *Only the female mosquitoes bite.* They don't have enough iron or protein to make their eggs, so they need to get them from an outside source — blood. (This, ironically, is also believed to be one of the reasons humans started eating meat — because we females needed the extra iron and protein to create our own babies.) Your blood goes right into a mosquito's stomach, gets digested, and is turned into eggs. In mosquitoes, only females who have already had one batch of eggs spread disease. Think about it for a minute . . . mosquitoes by themselves don't carry diseases, they pick them up from their sources of blood (us or other animals). And the only way they spread a disease is if they are going back to the "well" for the second time. In her lifetime, a female mosquito can't really lay more than three batches of eggs (up to five hundred eggs in total). So it's really a relatively small number of mosquitoes that cause disease. It's the older ladies, in fact. (This leads me to wonder if historically the male obsession with purity in women has to do with a fear of getting diseases. Perhaps. And yet males having multiple partners is most likely how human sexual diseases spread. But I digress . . .)

Taking a break from reading about mosquitoes, I decided to watch TV. Nerd that I am, my favorite TV streaming choices are almost always documentaries. I scrolled through the listings. Hmmm. I had never watched the PBS documentary about Rachel Carson. I had read *Silent Spring*, which many environmentalists I respect cite as the trigger for their awakening. Also I had received the Rachel Carson Award from the Audubon Society many years ago. I was surprised I hadn't watched this documentary before.

Welcome back to the magic of the Trail of Books (which is also the Trail of Movies and Documentaries). It's the magic that happens when I follow my instinct in choosing what to read or watch, and it turns out to be exactly what I need to learn about.

Yes, I remembered that Rachel Carson had written about DDT. But what a book doesn't capture that a documentary can are the actual sights and sounds from the times in which the author was writing. Including the deep, authoritative, WASP-y male voice proclaiming that through the miracle of chemistry and DDT, man has now *won* the war against the mosquito and will ERADICATE it from Earth, accompanied by black-and-white footage of a truck spraying clouds of white powder over kids eating sandwiches at a picnic table, and airplanes squirting white dust over fields and suburban neighborhoods while kids play outside. The HUBRIS is embarrassing to watch. Sure, mankind has eradicated a lot of species, mostly by accident or for the pleasure of blood sport — like the passenger pigeon, poor thing. That bird never bit, infected, or threatened anyone. In fact, it helped people by carrying messages from one place to another on our behalf. And was delicious to eat, reportedly. So how did we thank it? Kapow. Dead.

It's normal for many people to want to be a hero, to want to save the day. Humans strive to eliminate enemies and control the world in order to protect their progeny from disease and death, and even more importantly, to protect and expand their wealth. But killing things (other people, pests, alleged enemies) doesn't solve our problems. As illustrated in the documentary and a thousand scientific studies, when we try to kill things like insects or weeds or viruses and bacteria at a grand scale, they develop resistance and rise up even stronger than before, requiring even greater interventions and stronger chemicals to ramp up the fight. As a by-product of our attempts at insecticidal and bactericidal genocide, we also kill many other things that we need and love, like butterflies and bees, birds, good bacteria essential to our health, and all the amphibians. This leads to a chemical, genetic, or military escalation in which we end up killing . . . ourselves. We end up killing that which we LOVE.

Ahh. Now I get it.

Mosquitoes are an integral part to a much greater food chain. Male mosquitoes don't bite and suck blood but eat plant nectar; they are pollinators. Mosquito larvae are food for dragonflies, fish, birds, some toads and frogs, turtles, spiders, and ants. Those water-dwelling babies (larvae) also eat algae and bacteria, which helps keep water clean. *They keep our water clean.* That's important!

What, then, is the answer to the "problem" of mosquitoes and diseases? Well, as my grandfather liked to say, *prevention*. On a small scale, preventing mosquito breeding is about reducing standing water such as in old tires, pots, and even the nooks of plants like the bromeliads. On a larger scale, in places like Africa and South America where fatal mosquito-borne diseases are serious threats, especially for children, it's about installing plumbing and providing clean water, secure housing, window screens, and medical assistance, including vaccines. Mosquito netting over beds is nice, but it turns out that most mosquito bites occur during the day. We need to find ways to prevent and heal the mosquito-borne diseases already inside *us* so that a female mosquito bite does not result in *her* getting infected.

Genetically modified mosquitoes (sterile male mosquitoes) have been released into the wild in Florida. This is *not* an optimal solution. If this technique successfully eliminates mosquito populations, the results would be catastrophic for insects, birds, and animals that feed on mosquitoes, and water quality would drastically decline.

Humans must commit to creating an environment in balance with nature, where nature's food chain is allowed to proceed naturally. How do we do that? I think about this question a lot. It can't be done by eliminating populations of "pests."

Some people have argued that we need to limit human population growth in order to live in balance with nature. In the past, humans have come up with horrible methods to "control the population" of "unwanted groups." Proponents of eugenics murdered, imprisoned, and secretly sterilized the groups they deemed undesirable, leading to horrid traumas and terrible suffering. There is absolutely no situation where this sort of behavior is acceptable, even though it still happens today.

What's a more productive way to create balance? Education. Especially the education of girls and women. When girls and women are educated, they understand how their bodies work. If they live in a culture that allows it, they can decide for themselves how many children they want to have, and then care for those children with a greater amount of attention and ability to support them. They are also more likely to take on leadership roles in government and businesses that protect and nurture the healthy development of future children (which is why education for girls is so threatening in highly patriarchal cultures). But the education of boys and men is essential, too. Especially teaching them to learn how to communicate and connect with others in positive, constructive ways — in particular with girls and women, who are vital to their survival. The global human population could decline naturally over time, providing more breathing space for humans and nature to live in harmony with each other.

Now, before the capitalists among you freak out about population decline because capitalism is a system that requires continual growth (I'm not naming names, but you know who you are), read on. These changes need to be accompanied by a new economic model that doesn't rely on endless growth and the domestication (aka servitude) of women to do all the household duties. This would require not just education but radical culture change — in almost every culture around the world. Men will have to pull their domestic weight — or pay real money for others to do it for them. In my view, it's fear of this kind of change that is fueling a backlash against feminism right now, everywhere from Texas to South Korea.

Some men I know fear population decline because of the potential loss of our civilization, or human consciousness. But the more I journey, the more I realize that consciousness cannot be destroyed. It exists with or without our bodies. And every civilization and culture evolves and changes over time. It's only natural!

People need to learn how to live in balance with nature. We also need to learn to live in balance with *each other*. When men and women (and everyone else) are *truly* free, educated, and loved, we are all better able to take care of the world around us – and enjoy it more!

It's very possible that we don't need to reduce our population to live in harmony with nature. But in that case education is even more essential because we must learn to create, invent, and innovate new ways of living on this very special Earth.

If we humans don't commit to being good stewards of the environment around us – even if we don't like what that demands of us – nature will take care of the environment in its own way. And the mosquito will be just fine. Fabulous, in fact. Meanwhile, we humans will be decimated by more and more diseases, spread not by just mosquitoes, but by our own ignorance.

After watching the Rachel Carson documentary, I found the courage to watch another one, this one aptly called *Mosquito* (because, obviously). It was a depressing experience, and afterward I realized I needed to go back and talk to Mosquito again, and to Bat too. After all, bats and mosquitoes are partners – predator and prey.

I decided I would journey the next morning. In the middle of the night, I woke up with these words in my head: "It's not really the mosquito that's the vector of disease, it's us. Our travels. Our trash. Our toxins. Our tragic belief that killing something will make it go away." I emailed that message to myself and went back to sleep.

One of the most depressing stories from the mosquito documentary was about mosquitoes breeding in abandoned tires. What I didn't know before watching *Mosquito* is that old tires are shipped all around the world. If mosquitoes lay eggs in water trapped in a tire, and then that water dries up, those mosquito eggs are still capable of hatching once they get wet again. Even if the tire has been moved to a different continent in the meantime. The deadliest of the mosquitoes are spreading everywhere in junk tires. And because climate change is causing overall warming, mosquitoes can survive farther north, and their range of influence is spreading more and more. Car tires. Fucking car tires.

The next morning, I began my return journey.

✳ ✳ ✳

I immediately entered the mosquito mothership and apologized for my ignorance on my previous visit.

"How can we learn to peacefully coexist with each other?" I asked.

"Ahh, now you are asking the right question," Mosquito said. "You must heal the diseases within *you*. Stop killing our natural predators. Take care of your own people. Lift them out of poverty. Clean up the squalor. And stop heating up the planet. If that continues, mosquitoes will be the least of your worries." This time she spoke with kindness. I felt humbled and grateful to her. We even hugged. It was awkward, but still. It was a hug.

"Now, go talk to Bat." She dismissed me.

I went back to the bat cave and the cavewoman was there. "Oh, not you again!" I said. The last time I had seen her, she had murdered me and eaten my heart. She ignored my comment and handed me the same leaf mixture I had eaten on my previous visit. I ate it, and I found myself with Bat. I had to wake her up.

"I'm sorry I didn't ask you before, but what is your job and your purpose?"

"Our job is to keep nature in balance."

"How can I help you?" I asked.

"Tell people to leave us alone — even the scientists and researchers! They need to respect our homes. We wouldn't do to them what they do to us. Tell them to get their heads out of their goddamned test tubes. You can't understand anything unless you look at the whole thing. The web of life is *real*."

I thanked her. The drum was still drumming, and I wasn't sure what to do next. Then I found myself on the electrical grid or web that I had flown through during my previous bat journey. I started jumping on the web as if it were a trampoline. I was having a bit of fun, but suddenly a bat swooped in and ate me. Now what? I asked myself. Before you know it, she pooped me out right on my side porch, where all the other bat poop is. She had brought me home. The journey was over.

* * *

Thinking about this journey, I begin to realize that once I journey to speak with a being that I am initially afraid of, I start to feel genuine affection and love for them. And that happens because we learn to trust and know each other. What was once an annoyance becomes a real friend. What I once wanted to eliminate, I learn to appreciate — I realize these beings really aren't so bad after all. They are not my enemy. They have feelings too. Indeed, the feeling I carry out of a journey is love. I feel my heart softening and warming in ways I never could have made happen through an intellectual analysis of a creature's role in nature — or even from watching a documentary.

And this, my friends, is why journeying is such a valuable tool. Doing research doesn't shift my heart. It is only through a personal relationship of trust that I can learn to love those things that once were just annoyances. I can assure you, these experiences have changed how I behave in my day-to-day life, too, for which I am grateful. If anything, journeying makes me want to learn even more about these creatures — but not through the kind of research that stomps into a bat cave without regard for the feelings and rights of whatever lives in there. We must learn to respect each other's homes and families. We must stop killing what we ourselves love. And we must even stop killing the things we don't love. Because *somebody somewhere* loves them.

I have no doubt that 236 million years from now mosquitoes will still be here on Earth. Us? Probably not. I mean, how many tires will there be on Earth in 236 million years if we don't change our way of living? A lot.* And no, sending them out into space, like Jeff Bezos thinks we should, won't solve our problem. Neither will

---

* I have since learned that particle pollution from car tires is thousands of times worse than car exhaust emissions. Someone really needs to invent a new way to get around. Please.

sending them to Mars. The web of life is out there too. We just haven't learned to see it yet.

I think I'm going to leave my moss-covered pile of tires exactly where it is as a reminder that all our actions have consequences. And that to stop killing what we love, we have to stop trying to eliminate those things we don't easily like.

Thank goodness for anti-itch creams!

Thank you, Mosquito. And again, thank you, Bat.

Tulip Tree

Travel the path of love.

My house is nestled between two very tall trees, which I have known as tulip poplars. They really are huge, about the height of a seven-story building. Despite their name, tulip poplars are not very closely related to poplar trees, although they are fast growing like poplars. The bright green leaves are shaped like tulips, and in the spring the branches are covered in yellow-and-orange flowers that look like tulips as well. *Liriodendron tulipifera* goes by many common names that include the word *tulip*, but it's also known as white wood, fiddle tree, and yellow poplar.

The Leni Lenape used these trees to build canoes, since the trunks are tall and straight and thick and strong. Early European settlers called them canoe trees. But I have no need for a canoe, so I will simply let them grow and see how tall they become while I am still alive. They are the tallest Eastern hardwood tree.

I was delighted to discover that tulip poplars are the wood of choice for organs. I love a giant organ played with gusto. But still, I won't be chopping these trees down for any reason. They are too beautiful.

They do, however, spread their seeds everywhere. I must have pulled out at least a thousand baby tulip trees in my garden. Some of them not so babyish, either.

After the mosquito journey, I needed to take a break and journey to something that felt more benign. And lately I had been noticing how my two giant tulip trees seemed to stand sentry on either side of my house, front and back. Hmmm . . . I wondered if that signified something.

One of the things I love most about journeying is how surprising the experience can be. I never know exactly where I will go and what I will learn and see.

* * *

As I settled in for this journey, I could feel my power animal circling around me like a dog that wants to go outside (but my power animal is *definitely* not a dog). I was happy to follow them down a dark path into the earth. (Sometimes she is a she and sometimes he is a he — they are beyond gender.)

At the bottom of the path was a dark cave where some Indigenous men and women in traditional dress, with beading and feathers, sat in a circle around a fire. They stood to welcome me, and I bowed to them in greeting. They smudged me. Then they passed around a sacred pipe. I sat patiently waiting to hear what they would say. A man began to speak.

"We thought we could trust the white people when they first came. We were seduced by their colorful beads, drink, and fire sticks. We were no angels either — we had our own wars, our own history, our own stories. Perhaps we were brought together to learn how to transform and transcend — to realize that we are all one tribe."

He called me Little Sparrow, which made me cry because I understand what a gift and honor it is to receive a name from an Indigenous person.

We sat around the fire and held hands. As we did, the fire grew taller and took on the shape of a giant tulip poplar.

"The tulip trees are your grandfather and grandmother. They are there to protect you and keep you safe. When you love nature, you begin to walk the path of love, and that is what we are here to do. There is no evil in nature."

"What about evil in people?" I asked.

"Many people do very bad things, but that is the path they have chosen to walk. You must learn to protect yourself. There is as much deceit in the New Age movement as there is in all the churches."

"How do I know what or whom to trust?" I asked.

"First, trust your heart and intuition. And then, you can always trust nature. Nature may get angry and rage. Nature can be sorrowful and dangerous. But nature is never evil. Our job here is together — no matter our color, our race, our religion, or our gender — to learn to walk the path of love in harmony with nature. And to realize we are all one tribe on this Earth."

He continued, "We, like the tulip poplar, cannot be destroyed. We can be cut down or killed, but our spirits live forever. The sorrow of death is real, but temporary. We are here to learn how to love — each other, nature, and the Great Spirit."

I then found myself alone out on a lake in the dark in a hollowed-out log canoe. It was a stunning night, a palette of dark blues of the night sky and dark greens of the trees silhouetted on shore. The water was a mirror reflecting the stars. I lay back in the canoe, staring up into the vast universe. There, looking down on me with love, were the star images of a grandmother and grandfather. I thanked them. I thanked the people I had spoken with, and I thanked my power animal. The journey ended, and I felt filled with love and gratitude.

✶ ✶ ✶

The first thing I did after writing down everything I remembered from this journey was to look up the meaning of the sparrow in a book called *Animal Speak* by Ted Andrews. It said, "The sparrow has not always been considered the pest it is today. It is a perky, assertive bird that can hold its own against many forms of predation."

Yup. That sounds about right.

Do you remember that the architect who worked with me to design my house said he was a shaman? His name is Michael Jonn. Years after our house was completed, I asked him to recommend a book on shamanism to me. He recommended Ted Andrews.

Ted Andrews was a teacher, mystic, and the author of several books, including *Nature-Speak* and *Animal Speak*, which are incredible sources for understanding the history, symbolism, and meanings of plants, animals, insects, and other creatures. His books are meant to be used as tools to help you understand and listen to what nature might be trying to communicate to you. Those two books started me on my path and opened me up to observing, listening, and recognizing that everything in nature is important. *Nature-Speak* oracle cards helped guide me through some of my hardest moments and taught me to trust my own intuition and heart.

To my point, I felt that Andrews's description of the sparrow was very apt for me. I started to pay attention to other references to sparrows. Especially little ones. (Dolly Parton, I'm talking to you.)

A few weeks later I came across a fascinating video on TikTok about Mao Zedong's war on sparrows, which I had never heard of before. In 1955 Chairman Mao, the leader of Communist China, set a goal of China becoming self-sufficient in agriculture through the collective industrial farming model set up by the Communist Soviet Union.

In 1958 Mao launched the Four Pests campaign to eradicate flies, mosquitoes, rats, and sparrows as part of what was called the Great Leap Forward. The story goes Mao was visiting a farm and saw a sparrow pecking on some rice. Upon Mao's single statement declaring the sparrow detrimental to agriculture in China, the people "rose up" against sparrows and in the next five years killed ALL OF THEM. Seriously. All of them.

But what happened next wasn't what they had expected. Turns out, sparrows weren't eating the rice, they were eating the grasshoppers and locusts. This gross error, along with a major drought and typhoon, led to the Great Famine in China. The campaign (which also included the collectivization of all farmland and the use of a great deal of agricultural chemicals) was an utter failure. After the campaign, China had to import 250,000 sparrows . . . from the Soviet Union.

Mao's Great Leap Forward was truly a Leap of Death and Destruction.

Human stupidity — usually driven by ego, hubris, and greed, and even a false sense of patriotism (the Chinese people *thought* they were doing the right thing for their country) — seems to know no bounds. I continue to be shocked by what I learn about all our mistakes. (I'm deeply grateful for the internet and social media, because it provides access to an incredible range of information that was not easily accessible before.) It is estimated that the Four Pests campaign combined with the Cultural Revolution (from 1966 to 1976) resulted in the deaths of as many as twenty million people in China. This is a reminder that all races, all religions, all ethnicities, all political parties, and all nations are capable of horrible, brutal behaviors. It is those behaviors that we must try to understand and heal, rather than continuing to pursue eradication campaigns against some group of living beings that currently annoy us — whether they are human, bird, insect, weed, or animal.

Ironically, in 1973, during the final days of the Cultural Revolution in which "intellectuals," "scientists," and "elites" were shot in the streets, my father was part of the first group of journalists invited to travel to Communist China. He came back with a very romantic view of Chinese agriculture and Chinese society. Propaganda works. That does *not* mean we should demonize Chinese people. We have *all* done stupid things. And we are *all* deserving of compassion.

I am not a communist and find it an abhorrent concept. Trying to control people and make everything fair and equal decapitates our human desire for personal expression and fulfillment, joy, and love. That doesn't mean I believe capitalism (or socialism) will get us to where we want to go either. As I like to remind people, Adam Smith, the primary founder of capitalism, lived with his mom, who cooked all his meals, did his laundry, and kept the house clean "for free." We *really* need a new economic model. The current economic theory (value for shareholders over value for all, with no valuation of nature) is a drive for perpetual growth, profits, and wealth. It is human nature to strive for dominance, for self-actualization, for freedom. Our deepest need is the freedom to *create*, whether it's music, art, businesses, homes, families, landscapes, or ideas. But nature is never in perpetual growth — nature is made up of cycles and waves. Seasons. Lunar cycles. Days and nights. Sleep and waking. Birth and death. The more we can surf the cycles of nature, the more we can create magic. And love.

It is only by traveling the path of love that we arrive at someplace worth going to. And love values *everyone and everything*.

A few days after the journey, I was out collecting kindling and decided to pay a visit to the tulip tree behind my house. It's the tree I gaze on through the window whenever I am taking a bath, so I know it well from a distance. I walked up closer to the tree and saw it as if for the very first time. At first, it seems not just one tree but two. But they are joined together at the base and also a few feet up from the ground, as if they are kissing. I was both happy and sad when I looked at it. That tree is what I dream true love is like — together but also different. United, but both able to reach for the sky independently. Perhaps one day I will find something like it. Or perhaps I won't. Perhaps those two tree trunks represent the male and female inside me, balanced, intertwined, content.

Either way, I am happy to continue walking down this strange magical path of love and see where it leads, knowing that my star ancestors are watching over and protecting me with *their* love.

Thank you, Tulip Tree.

Thistle

Dig deep.

Most gardeners hate thistles. They have sharp, spiny leaves and stems. The purple flowers are pretty, but they turn into fluffy messes when they go to seed. I, too, have been known to deeply resent thistles. I also have been known to pull thistles with my bare hands. I'm telling you this not to brag, but to warn you against trying it. (Although let's be honest, I'll keep doing it anyway.) More than once, I've been woken up in the middle of the night by a throbbing in a finger that demands immediate attention. It's a thistle splinter. Each is like a tiny shard of glass, so fine and transparent that I must find my magnifying glasses and seek the bright fluorescent lights in my kitchen pantry closet (also known as my doctor's office) to gouge it out so I can go back to sleep, for Lord's sake.

The sharp, spiny, pokey, fuzzy spines that can inflict terrible pain protect thistles from being eaten by animals. Yes, the plant is edible, hence why it needs protection. But most reports state it's not worth the effort to try to eat thistles. There are about sixty thistle species that range from highly adored, like the artichoke, to those considered noxious weeds by farmers. In fact, thistles are very beneficial. The flowers are hugely popular with pollinators (including monarchs and many other butterflies). The seeds are an essential source of food for birds, especially the gorgeous goldfinch. And the fluff that flies all over the place when the seeds release from the plant are prime bird nesting materials. If I were a bird, I would want a nest made from that fluff. It's like the cashmere sweater of nests.

But I am not a bird. And I grew up with the frequent reminder of "one year's seed, seven years' weeds," so within the confines of my backyard I have tried to eliminate the thistle, which I know is a game that can never be won. Its roots can grow fifteen feet deep and spread up to fifteen feet wide. When I put a big stone on top of a thistle to block it, the thistle just springs up around it. What does work is to plant slightly more aggressive species to outcompete it. For instance, I had a bad patch of thistle in a corner of my garden, so I planted horseradish, which has a hefty, deep taproot, and pineapple mint, which is about as aggressive a plant as you can find. That corner of the garden now looks beautiful, if I say so myself. And if I need some fresh horseradish, I know where to find it.

My journey to Thistle was very brief and to the point (ouch). I did the journey in November, when all the life had returned to the roots for the season, so I went underground. I had a feeling that's where the message would be.

* * *

I followed the roots down deep into the earth. And then Thistle spoke.

"We dig deep to bring things to the surface to share with the birds and the bees. Dig deep, dig deep," she repeated over and over. "Leave us some space to bring things to the surface to share. Our flowers feed our friends, and your friends. And then our children, our seeds, fly off on their own, and we let them go free while we dig deep with our roots."

I saw a blue sky filled with swirling seed fluffs and watched them fly off to parts unknown. And that was it.

* * *

Like I said, it was a very brief journey. Yet, as with many other journeys, I came away with a shift in perspective, moving from skeptical and annoyed to impressed, grateful, and filled with love. Suddenly, surprisingly, I want more thistles in my garden. OK, maybe that's an exaggeration. But I will no longer label thistle as noxious or a pest. After all, I love the birds that eat its seeds and the butterflies and bees that drink its nectar. And the swirling fluffs sailing off to their own adventures is something that children find magical. Maybe it still is magical, even when we've grown up. What harm is there in letting thistles grow in the wild? And what might we learn if we take time to dig deep into our own lives?

Dig deep (1 foot). My first thought was about letting our children go off on their own while getting ourselves rooted and digging deep to discover our own gifts. Not enough parents do this. Too many parents keep a firm grip on the lives and dreams of their children. My parents did that to me (hello, family business). We are not our parents. We are not our parents' parents. And yet we are all seeds of their seeds.

Dig deep (2 feet). As I dig deeper into the thistle story, I find treasures. For example, milk thistle, *Silybum marianum*. (Say that out loud for a smile.) It is also called Our Lady's thistle, holy thistle, or St. Mary's thistle, after the Virgin Mary. The milky fluid that comes out of the stem when milk thistle is picked is supposedly evocative of Mary's milk that fed the baby Jesus.

Dig deep (3 feet). While some believe thistle to have healing properties — especially for the liver and digestion — it's not universally accepted or scientifically tested. My mother injected herself daily with milk thistle when she had breast cancer, but it didn't heal her.

Dig deep (4 feet). I turn to Ted Andrews and *Nature-Speak*. What does he have to say about thistle?

> *Are we being too defensive? Are we exposed to constant complaining and criticism? Are we doing everything we can to help ourselves? Do we need to clean up some aspect of our life? Thistle reminds us to maintain pride in who we are and not be afraid to*

defend ourselves. There is such a thing as righteous anger, and
we do have the right to defend ourselves from the criticisms of
others. When thistle shows up, it is time to do so.

Interesting. I don't know if Ted Andrews is correct, but these are always good questions to ask yourself and consider.

Dig deeper (5 feet). The thistle is part of the sunflower or daisy family, and birds and bees enjoy many plants in this family immensely. Funny how plants have families too. I wonder how whoever it was that decided to put the sunflower and the thistle in the same family made that decision. Do thistles and sunflowers get along nicely with each other? I've never seen them fight. Thistles are sharper. But sunflowers get taller. They probably get along nicely.

Dig deeper (6 feet). I think back to the part of the journey where the fluffy seeds — Thistle children — danced away in the wind. Do you look for reflected glory from your children or partner? I think the term "reflected glory" reveals a lot about our human nature. I see it when people choose relationships because of how the connections will shine back onto them. When someone is beautiful or successful, sometimes we want to be near them because it makes us look or feel better about ourselves. But are we developing our own dreams rooted in our own identity? Are we connecting with others based on who we really are deep down inside? Those authentic connections with others are more long lasting and *actually* make us feel better about who we really are. Sometimes I think we need to learn to see with our hearts, not our eyes. How often do we judge and value people based on surface appearances, making assumptions based on how they look or act without trying to understand what's underneath? That's a trick question, and there is only one answer: too often. Thistle can relate.

Dig deeper (7 feet). Some types of thistles are "imperfectly dioecious," meaning that a single plant bears both male and female flowers, sometimes. Depending. It strikes me that lately humans have become obsessed with defining gender and sexual orientation. But nature is super gender fluid. *And* super sexual. Perhaps humans are finally simply allowing themselves to dig deep enough to discover their true nature?

Dig deeper (8 feet). What have you buried deeply that needs to be brought to the surface? Often it's a secret longing and sexual desire. It's my belief that the more society, governments, and religions try to control people's sexual desires, the more those desires go underground and create destructive forces in innocent people's lives. All too often it's the most vocal figures of moral authority who get caught "with their pants down." Sexual desire is normal, healthy, and wonderful — if we allow it to be expressed freely, with love and consent, and without harming others.

Dig deeper (9 feet). I heard a true story from a farmer friend of mine about a Mennonite farmer who went to visit his father on his deathbed. The father was old and had lived a good life. But as he lay dying, he said to his son: "Remember that farm we had years ago, along the river?"

"Yes, I remember it," the son said.

"Remember how I made you and your brother pull thistles all summer?" the father asked.

"Yes, I remember," the son said.

"You know, it didn't do a bit of good," the father said.

Dig deeper (10 feet). It's cold and dark down here. Quiet, too.

Dig deeper (11 feet). No, wait, what is that sound? It's like a humming, chomping, swirling. It's the sound of the life of the earth and soil and rocks breathing, consuming, living.

Dig deeper (12 feet). Sometimes, late at night if I can't sleep, I try to imagine all the things happening on planet Earth all at once in that moment. All the sex. All the births. All the deaths. All the tragedies.

All the joys. All the eating. All the defecating. When you add nature into that mix, it's one giant ball of swirling, orgiastic, pulsating activity. Visualizing everything all at once puts any paltry worries I might have into perspective.

Dig deeper (13 feet). I feel safe down here.

Dig deeper (14 feet). It's getting warmer now.

Dig deeper (15 feet). Ah, I can hear it. The heartbeat of Earth. It sounds like love.

Thank you, Thistle.

Deer

Suffering begets suffering.

JOURNEY: NOVEMBER 17, 2021

When we built our house in the woods, I knew immediately that I wanted to — no, *needed* to — surround the backyard with a deer fence. Otherwise I would never be able to garden in peace. I had heard enough complaints from gardeners about deer my whole life to know I had to be proactive. And only once have I looked out the window and seen deer grazing in my backyard. A tree had fallen on the fence, and they had jumped in. I chased them out and fixed the fence, posthaste.

My family and I quickly learned that we had moved to prime hunting territory. We got dozens of requests for access to hunt for deer on our property. (Still do.) People even offered to pay us thousands of dollars for access. We finally settled into a relationship with a local family of bowhunters (no guns allowed). In exchange for access to hunt, they plow our driveway when it snows, help keep the trails clear, and occasionally bring me a deer to eat. My first time tasting the meat from a deer that had grazed on my land was a spiritual experience. I felt connected to the cycle of life — and *nourished* — in a way I had never experienced before.

People's relationships with animals are rife with turmoil. Especially when it comes to what we choose to eat or not eat. Talking with people about food choices has become as controversial as discussing political or religious beliefs. And it seems that people are all willing to tell you what is "wrong" with the way you eat. Often, *very* meanly. (Or so I've found.)

There is also a dichotomy between people who want to save the animals and those who want eat the animals. But it's usually a divide based on incomplete information and a romantic ideal of our relationship with animals.

Take kangaroos, for example. Adorable! When wildfires ripped through Australia in 2019, headlines across the world expressed sadness about all the animals being killed (the iconic image of a kangaroo silhouetted against a burning background became the symbol of the tragedy). It is sad. It's terrible. Tragic. But *every year* hundreds of thousands of kangaroos are culled because they are invading suburban neighborhoods and overeating everything out in the bush — and they lack enough natural predators (dingoes) to keep populations under control. In fact, you can buy kangaroo testicle bottle openers at airport gift stores in Australia — and, of course, on Amazon. (Full disclosure, I bought one while in the Outback. As a gag gift.) Instead of simply culling the kangaroos and selling the hides for leather and the meat for pet food, people could eat them and save *millions* of acres from being overgrazed by cattle and sheep, which are not native to Australia.

Lucia and I were in Australia for Christmas and New Year's in 2019, and she was appalled that I ate kangaroo. I don't blame her. It's normal to be appalled by the idea of eating something cute. But cows are cute. So are pigs. I have Aussie friends who lament the fact that they should eat kangaroo more often, but their kids are reluctant because of the taste. I hardly taste any difference between kangaroo and beef. And the more kangaroo, buffalo, and venison I eat, the more beef tastes strange to me . . . almost rancid. Speaking of rancid, most of the "fresh" grass-fed organic beef sold in the major supermarket chains in the United States comes from either Australia or South America. (At least that's what it says on the label.) By the time it gets to you, it's been in that plastic packaging for quite a long time. It's an unfortunate aspect of a food system that encourages consumers to expect a consistent supply of any type of food they desire, at the price they expect, all year round. Scientists discover that meat from grass-fed cattle is healthier for humans to consume than meat from cattle raised on grain (which makes sense since cattle's stomachs are meant to eat grass). The media writes about that discovery, and then consumers start searching for grass-fed beef. Then supermarkets feel pressured to deliver it to customers in a consistent, year-round way — even if that requires shipping it from around the world.

Instead, find a local farmer who offers organic *seasonal* (preferably regenerative organic) grass-fed meat, if you can. Buy directly from a local farmer and bypass the whole industrial meat process. (Farming is hard work, and I have tons of respect for people who grow food and keep humanity fed, including those farmers in Australia and South America. But we have to find better ways to do it that doesn't destroy the Earth in the process.)

Honestly, it's a bit idiotic. But we humans have been known to be idiots. (If it makes you feel better, I've spent my whole life studying this stuff and I still often get confused.)

The reason buffalo no longer roam the prairies of the Midwest and West (except where they have been reintroduced) is because during the nineteenth century, white men killed as many of them as possible in a fun sporty effort to starve the Indigenous people of the region. Sometimes cruelty is the point.

When Temple Grandin spoke at that conference in California that helped start me on my shamanic path, the thing that struck me most was her discovery that many people who worked at factory farms and meat-processing plants actually *enjoyed* hurting animals. Though she is an expert at designing cruelty-free livestock processing systems, that joy from cruelty is a flaw she could not engineer out of the system, as gifted as she is at understanding how to help animals suffer less. I have often thought that when we eat meat from animals that were raised in suffering, we ingest that trauma.

I have always been a pet and animal lover. The farm I grew up on had sheep, steer, pigs, chickens, geese, turkeys, and ducks. We ate them. And yes, I eat all the meats. As a publisher and writer I have witnessed an insane amount of arguing over what to eat and what not to eat. Between health concerns and environmental concerns — between the plant-based eaters and the meat eaters — it seems that no one can agree on anything and every day there is a new dietary craze. (Dietary crazes always sell well, which is why people keep inventing them.)

Meanwhile, as we continue to argue, animals are being tortured in factory farms. Farmers are held captive by the giant meat-processing corporations. Slaughterhouse workers are being abused and under-paid. Toxic chemicals are being fed to farm animals to fatten them up without regard to animal welfare so industrial-scale food companies can make higher profit margins. Factory-farm animal waste turns into toxic methane waste rather than precious fertilizers for farm fields. And everybody complains about the cost of food (and their health problems) and blames it on "the government." (The govern-ment is us and only us. The government is *you*.)

Yet at the same time, deer eat for free, from nature. Deer poop in the woods, nourishing the soil. And many people are willing to pay lots of money to hunt them for fun. It's crazy.

Venison is delicious. Indigenous Americans understood that deer are an essential part of the joy and pleasure of eating and living. They used every part of the animal for some good purpose. For thousands of years, Indigenous people around the world were dependent on their local protein source and thrived because of it — thanks to deer, buffalo, kangaroo, salmon, whales, and seal. I feel infinite gratitude and respect for those animals and fishes that have sustained human-ity from the beginning of our time on this planet.

Sadly, most of the hunters I know are on the part of the political spectrum where debates about food and environment aren't hap-pening, unless it's about the cost of food (they want it cheap). Many hunters even spray Roundup in the woods to kill the woody plants and encourage meadows where deer like to feed. (It's easier to hunt them that way.) Some even plant GMO corn for the deer to eat. (But not in my woods!) Sadly, many deer now suffer from chronic wasting disease due to overcrowding, and also being fed by humans. The guys who hunt my land say adamantly: "Don't feed the deer!"

I have been enjoying watching the deer families graze in my front yard for more than eighteen years. Each spring, a new batch of adorable fawns appear. Usually there are three fawns, surrounded by their mothers and aunties and the occasional buck daddy or brother. I was very eager to finally get to meet with them in a journey.

* * *

The journey began when a door appeared in the tree. I opened it and stepped into a lush green meadow filled with deer romping and grazing. I felt a tug on my clothing. Looking down, I discovered I was wearing a white dress and a little fawn had grabbed the fabric in her mouth, pulling me toward a doe and buck who were clearly the elders. We bowed to each other in greeting and gratitude.

"We have been waiting a very long time for you to come and visit with us. Thank you for providing us with such a wonderful home," Doe said. I asked if they minded the hunters.

"Oh, no. It's our gift to you in thanks for your protection. We know that if there are too many of us there won't be enough for any of us to eat and we will all go hungry."

At that point I was feeling very agitated. I had been a little bit hungry when I started the journey, but now I felt like I was shaking with hunger. "Why am I so agitated?" I asked.

"When we are hungry, we are agitated. Hunger creates suffering, and suffering begets suffering." I started to cry, and they bowed their head to acknowledge the suffering that so many animals face when they are raised in confinement and squalor. Suffering begets suffering. I cried for all the humans who were hungry and acted out because they just needed something warm and good to eat. Suffering begets suffering.

Three years earlier, I had noticed that there was no longer any organic lamb available at my local supermarket. When I asked, I was told they sourced all their lamb from Australia and flooding there had wiped out all the sheep farms. I stopped buying lamb. Then, just a few miles from my house, a little sign popped up at the end of a driveway that said *lamb, goat, eggs*. I could see the sheep reclining in a green lush meadow (kind of like the one in my deer journey — which I am still in, by the way). Here was a farmer offering fresh, local, organic lamb. "Like that guy," Doe said to me, showing me an image of that farmer and his farm. "That's the right way to raise animals to eat. With love and joy and locally."

Doe and Buck handed me a fresh deer heart to eat. I am not one to eat organ meats, especially raw ones. But I knew this was a journey and I wouldn't be grossed out, and this was an honor the deer were showing me. So I ate it. In return, I handed both of them bouquets of plants from my garden (inside the deer fence), and they munched on them in ecstasy.

I then asked Buck about why human men are so obsessed with hunting deer and mounting their heads.

"Those men seek our dignity, strength, and power," he said. "But killing us and showing off our heads doesn't give it to them. Our masculinity comes from protecting our families, but also from the freedom we give each other. You can't own happiness, love, and joy. You can only find it in freedom. But not the freedom that says there is only one kind of freedom. You can only find it in the freedom to be wild, and yet still be protecting and caring for one another."

I then asked what I could do to help them.

"We keep things in balance, but we need more wildness to thrive. In return, we nourish you. Now, go eat something!"

✳ ✳ ✳

I came out of the journey and ate breakfast, grateful to have finally spoken with the deer. They had felt so calm and gentle, kind, and warm. The gentle strength of Buck felt like what I dream true masculinity could be. And the sweet kindness of Doe made me feel loved. I have to say, animals and plants speak more sensibly to me than humans do. All the hucksterism to create, sell, and eat "plant-based" meats and other highly processed foods just makes me depressed. Most of them are such convoluted products, marketed to assuage a guilt about eating meat and a fear of environmental collapse that can only be fully solved through changing the whole system of how we grow and harvest our food, not by the invention of yet another kind of processed food. (I will not be surprised to get hate mail about this statement. I apologize in advance for all the people I have offended, especially my friends, whom I love. But I stand firm in my position.)

A deer is plant based. Deer eat plants and only plants. Same with cows. Sheep. Kangaroo. Bison. Deer have been feeding humans since the beginning of time (human time). It's about as perfect a food as there is. As a woman (me) who has borne three children, had three miscarriages, and endured more than 2,580 days of extremely heavy periods (that adds up to seven *years* of my life), I can tell you I need meat to survive. If the female mosquito needs human blood in order to bear children, it makes sense that I need animal blood in order to bear children. I believe our food choices are personal and must stay personal. With one exception: Whatever types of food we choose to eat, we should educate ourselves about *how* that food is grown and processed, because *how it's grown impacts all of us.*

The Rodale Institute has been conducting a scientific study, called the Farming Systems Trial, that has examined chemical-based farming systems versus regenerative organic farming for the past forty years. I've been watching the results of that trial unfold over time. At first the study focused on whether it was even possible to farm organically and make a profit. The answer was a resounding yes — especially during droughts and floods because organic soil absorbs and holds water more effectively than soil on conventional farms does (which significantly decreases erosion). Farming organically also requires fewer fossil fuel inputs, which lowers costs for farmers and reduces toxic waste in the environment.

Next the Rodale Institute researchers started to study the soil more closely, and they discovered that all the life in the organic soil — the bacteria, the mycorrhizal fungi, and other microscopic creatures — were actually building up the carbon in the soil and storing it there in the roots of the plants. The same could not be said for the soil on the conventional plots, which store carbon at a *much* lower rate. That led to the profound realization that transitioning conventional farms to organic and environmentally friendly practices could sequester massive amounts of carbon and mitigate climate change. It's a simple but significant way to actually reverse the carbon buildup in our atmosphere.

The researchers have now moved on to study whether or not the organically grown crops are more nutritious and better for humans to eat. The studies are ongoing and beginning to document that yes, regenerative organic foods are more nutritious in often surprising ways. Other scientific studies from other organizations have also shown that eating organic foods prevents pesticides and other toxins from getting into our bodies, where they can persist and contribute to serious health conditions, so our health does benefit.

But the real learning I have witnessed is that the longer an area of land is farmed regeneratively and without chemicals, the healthier the environment and community becomes for *everyone* (whether you eat organic or not) — the water is cleaner, the air is cleaner, the farmer and farm workers aren't exposed to toxins, and the economy becomes much more resilient and successful. Workers at all levels of the system become wealthier and healthier. And . . . the land *heals*. It comes alive and thrives.

The number one comment I hear from people when they visit the Rodale Institute is that it feels magical. That's because it is.

When we work in partnership with nature, we *all* benefit.

Getting back to deer, while doing research for this book I read an article on garlic mustard, one of the latest "invasive weeds," and the role of deer in the management of this plant. The classic approach to invasive weeds is dig them up and remove them (or just kill them, often with toxic herbicides). In the case of garlic mustard, this brings to mind the German word *verschlimmbessern*. It means the act of making something worse while trying to make it better.

Dr. Bernd Blossey, a conservation biologist at Cornell University, discovered after studying garlic mustard for ten years that the best strategy for controlling it was to simply leave it alone. Not only that, but there is a strange and still-mysterious connection among garlic mustard, earthworms, and deer. Bear with me here: When invasive nonnative earthworms infest an area, and at the same time there is an overabundance of deer, that area becomes the perfect habitat for garlic mustard to proliferate. Just trying to kill the garlic mustard in this situation actually makes it more resilient. If it's left alone, it "self-regulates" and doesn't spread further. Therefore, the best strategy to prevent garlic mustard from spreading is to harvest the deer *before* the garlic mustard invades an area. If deer were harvested at a level that maintained the population at five to seven animals per square mile, not only would garlic mustard not be a problem, but neither would Lyme disease, according to Dr. Paul Curtis, the New York State wildlife specialist at Cornell.

For some perspective, I have counted up to twelve deer just in my yard, which is thirty-nine acres. There are 640 acres in a square mile. And yes, I have garlic mustard.

And those invasive earthworms? Well, that's a journey for another day and perhaps my next book.

Balance stops the suffering. It's the cycle of life. Everything eats and everything plays a role in the cycle, including us. The more the cycle is in balance and harmonized, the more likely everyone gets something to eat and nothing goes to waste, and the less agitated we all are. And the more we raise our food with love and kindness, in harmony with all the other animals, without chemicals and cruelty, the more we feel nourished by the whole process.

So much of the history of war is the fight for resources like food, fuel, water, wealth, or even women. If we change our approach to cooperating and collaborating instead of dominating and destroying, everything would get better quickly.

Imagine if we shifted our energy away from fighting each other and toward *feeding each other*. We would have so much more time to create and play. Imagine if we stopped choosing high-tech processed solutions and shifted our attention toward creating kind and resilient nature-based localized systems. There will always be some people who try to make their billions — just like there will always be male deer or male kangaroos fighting each other for dominance within their herds (the benefit being more sex and progeny). It's not just human nature. It's *nature*. But as a human herd we can also steer our evolution in new, better directions that add to our resilience as a species rather than destroying it.

Every animal has a purpose and is happiest when fulfilling that purpose.

Every human has a purpose and is happiest when fulfilling that purpose.

None of our purposes include only reproducing. That just happens no matter what. That's the baseline of life on Earth. Having children doesn't make us human. *Creating makes us human.*

And the true magic and power come when we connect with each other — and with nature — and create with kindness and love. Only we can end the cycle of suffering that begets suffering.

Thank you, Deer.

Paper Wasp

Keep it simple.

JOURNEY: NOVEMBER 18, 2021

In the summer of 2019, some wasps started building a gray papery nest on my side porch near my outdoor dining table. It got bigger and bigger. As people stopped by, they offered to take it down for me. I said no, let's wait and see what happens. And the coolest thing happened: The wasps built the nest up against a window, so when my family and I were inside the house looking out, we had a view right into their nest. I could watch them working so hard, laying eggs, the babies hatching, the babies growing up and flying in and out and in and out. I determined that they were paper wasps, which are beneficial insects. They eat other bugs, including caterpillars that eat garden plants. By the time the nest was finished it was twenty-six inches long and eighteen inches wide.

What interested me most was seeing people's reactions as they spotted the nest. Usually their response was a jump of fear, a double take, and an immediate instinct to kill the wasps. Men, especially, were ready to "come to the rescue" and do the dangerous and dirty deed. Instead, I invited them inside to watch the activity. It was like having one of those ant farm terrariums. Everyone was awed by the sight of the inner workings of the wasp nest. And in the fall, one by one, the wasps disappeared.

By winter, all the wasps had gone to wherever wasps go, and the nest was empty. In the spring, it wasn't wasps who moved back in (they make a new nest every year) but a family of red house finches that hollowed out the paper shell, built their nest inside, and had a nice cozy, safe, and lovely place to raise their little family.

Me and nature, we are a team.

Although the wasps built their porch nest before I had the idea to write this book, it was one of the experiences that led me to want to communicate with nature on a deeper level. It felt like an invitation to try to understand what nature wanted me to know, and why people are so afraid of all of it — the bugs, the wildness, and anything that seems unfamiliar. When the time came, I knew I wanted to talk to the wasps and find out what they had to teach me. Two years after the wasps left (I can still see the outline of their nest on my window), on a cold November night, I journeyed to encounter Paper Wasp. I was worried the wasps might be hibernating or dead, had forgotten about me, and would be hard to reach, but I was wrong. Lisa explained to me that when we journey to talk to nature beings, we aren't talking to a single being, we are talking to the spirit of all of them. In other words, it is possible to journey to talk with beings from the past. (And maybe the future too.)

✳ ✳ ✳

As soon as I entered my tree I swooped out and was flying around a summer meadow, gathering pollen, nectar, and wood pulp to bring back to the nest. I felt joyously productive. I flew into the nest and presented my offerings to the queen. She thanked me. I then asked what paper wasps' purpose is.

"We are builders. We build for the joy of building, but we are not attached to what we've built because each year we start over. We enjoy our activity and the pleasure of our summer work, and then we enjoy the restful winter. It's simple, and that's how we like it. That's the secret to our longevity. We are ancient."

"What do we humans need to know?" I asked her.

"Keep it simple. You complicate everything. And complicated systems are too fragile to last. We are ancient, but humans are like dust in the wind. Here today, gone tomorrow. Unless you can learn to simplify your lives. Enjoy the work. Then rest."

✳︎ ✳︎ ✳︎

Ironically (or not), while the wasps were building their giant nest, I was building an addition to my home. It was not simple. In fact, I often say: It's hard to get to simple. Sometimes that phrase confuses people, and they think I'm insulting them (the idea that simple is insulting is an insult in itself). Here's how I see it. We humans like to complicate things. Add things. Expand. Do more. Invent. Make money. Get rich. Create drama. Get angry at people who are different than us. Sometimes (OK, most times) we overthink things. We get distracted by "bells and whistles" and whatever the sparkly new thing is. But what brings true pleasure and joy are the most authentic, simple, and real things. Think about a bowlful of delicious homemade soup made by someone who loves you versus a fancy, elaborate, culinary foam–filled feast. The feast might fill your mind and ego, but the soup will nourish your heart and stomach. (Don't get me wrong, I enjoy the occasional fancy meal.)

When it comes to building, "it's hard to get to simple" means having the courage to leave things out. To not add doodads and ornate decorations. But it can also apply to the guts and engineering of a house. My house is extremely complicated (it has three different types of solar systems, for example). With the new addition, I was able to simplify my electrical system — by adding even more solar and Tesla batteries. Sort of. It's still way too complicated. It's no paper wasp nest that is made from local materials, reused by the birds the following year, and then biodegrading to close the circle. (Although my electric bill for April 2022 was only $15. Solar works.)

One day, I hope, we can design "homes of the future" that aren't just about adding high-tech solutions like "smart appliances," which often *add* to our carbon footprint. I believe that our creative challenge is to build homes that contribute to our environments in positive ways, like my solar home contributes electricity to the power grid. I think that, intellectually, people are ready to acknowledge climate change is real. But in their hearts and bodies they can't fathom how to keep their homes comfortably warm or cool without relying on fossil fuel systems.

I recently heard the phrase "change by design or change by disaster." Either way, change will come. But will we be its architects or its victims? The phrase reminds me of ongoing attempts to create machines to pull carbon from the atmosphere to store it in a safe place. People are way too eager to look to this kind of technology for a solution to our problems when the solution is right here already. We are standing on it. It's remarkable how we have overlooked the role of soil organisms, fungi, plant roots, and the mycorrhizae that live on those roots when it comes to pulling carbon from the air and storing it underground in the soil, where it belongs. (My theory is that it's because people haven't figured out how to get rich doing this.) Hundreds of studies have shown that when the earth's natural carbon cycle is allowed to occur unimpeded, the soil, fungi, the roots of plants, and the microbiome store the carbon underground. No added technology required.

It's math and systems engineering, really. If we all sat down at a table and fearlessly redesigned the whole system, based on what was best for the coexistence of nature and humans, and without worrying about who would make money, we could create a map to the future with incredible creative (and financial) opportunities for everyone. The difference between this sort of system design, as opposed to corporate efficiency design, is that rather than maximizing for profits, we would maximize for environmental efficiency and *resilience*. Which is kind of like maximizing for joy. Imagine what we could do with our time if we didn't have to worry about our planet and all the species being destroyed.

All we have to do is stop interrupting nature's brilliant and utterly simple process. How do we stop interrupting the process? Stop deforestation and excessive cultivation of farm fields. Make sure farmers plant cover crops and don't leave the soil bare. Leave nature alone when we can. And lastly — and perhaps most importantly — stop using synthetic chemicals that kill the life in the soil. Yes, I'm talking about going organic. But not just for organic food — we also need organic lawns, organic parks, organic sports fields and golf courses, organic forests. Do you know many environmental conservation groups use chemical herbicides to control plants that have been labeled as invasive species? Stop it! *We* are the worst invasive species.

Here's another illustration of my point. Try doing a browser search for "paper wasps." When I do that, for each listing of an informative article about these amazing insects, there are a dozen sites that tell you how to kill them and remove their nests.

This is what I want you to know about paper wasps. Studies demonstrate that paper wasps are capable of logical reasoning. They are even more intelligent than bees. They are very beneficial. *You should never kill them.*

Each nest has multiple queens, not just one. The queens are highly selective about who they mate with and find some male wasps more handsome than others.

Paper wasps are most likely to be eaten by dragonflies, spiders, birds, and the occasional mammal. In other words, they are an important part of the food chain too.

They are considered *eusocial* insects. *Eu* is Greek for "good." What's good about how wasps socialize? They share responsibility when raising their young, even when it's not their own offspring. They live communally and cooperatively, and there is division of labor. Some eusocial insects like termites even grow their own food inside their nests.

In other words, we might have a lot to learn from studying paper wasps.

During the summer when I was graced with the wasps' presence on my patio and window, I frequently had family dinners and friend dinners at my outdoor table. The outdoor couch where I drank morning coffee and enjoyed afternoon naps was just five feet away from their nest. I gardened nearby. The construction workers built my addition nearby. And not once, all summer, was anyone stung.

Thank you, Paper Wasp.

Poison Ivy

Pay attention.

JOURNEY: NOVEMBER 29, 2021

O f all the plants that are hard to love, poison ivy might be the hardest.

I itched. And it hurt. I had done something stupid, but I didn't regret it. However, I had to live through the consequences.

Eve was living in a house with three other young people on Cape Cod. As a visiting mother, the first thing I noticed was the front steps and rocky retaining wall covered with poison ivy. I said something.

"All the parents say that," my daughter shrugged, unconcerned. I was only there for a day and was dressed for the beach — a bikini under a sleeveless dress, and my trusty Adidas comfort slides. I went straight to the hardware store in town and bought elbow-length gardening gloves and a weeding tool — the one that looks like a screwdriver with a forked tongue.

"Get me garbage bags and a bin to put them in," I said with command, "and also plastic shopping bags." My daughter emptied the kitchen trash can and brought it out with a few plastic bags. Once I start weeding, it's hard to stop. I pulled out not just the poison ivy but all the weeds. I put the poison ivy into the garbage bags, to be taken out with the trash. Two hours later I had filled *four* garbage bags with poison ivy and made a huge pile of other weeds in the driveway, including a small maple tree that had been growing in a place that absolutely no one would be happy about a few years from now (including the maple). I had started out with plastic bags layered over my gloves as added protection against the poison, but after the first twenty minutes I gave up on the bags and just started grabbing with my gloved hands. When I was done, I threw the gloves away too.

Afterward, I showered and rubbed myself with Tecnu cleanser to rinse off the possible poison. Reader, you have to understand, for me to step into a tiny shower in a house shared by four twenty-something kids took a kind of fortitude I don't usually have (yes, I can be rather spoiled). But I had no choice. It would take a day or two to find out whether I would get a rash and how bad it would be. Washing with Tecnu (or something similar) was an essential step in limiting my future suffering.

Two days later, it started. There was a serious rash on the inside of my right elbow, and other bits and pieces all over my body. All over. None on my face, though, because I had used the tip of a shovel handle to keep pushing up my glasses while I worked. Thank the lord for anti-itch creams and potions. But I survived. And all week I couldn't help but wonder . . .

Who is poison ivy and why does she exist? And why do I assume she is a she?

My kids will tell you I'm a little bit obsessed with the summer camp I went to as a child. Camp Hagan was in the Poconos, on the Delaware River, and had to be abandoned in the 1970s due to a plan to build a dam (that never got built). I returned to the area many times over the decades to search for it, with no success. Finally, one day, I asked a park ranger and with their guidance was able to find it.

I am telling you about this camp because it is the location of the imaginary tree I often enter at the start of a journey. The camp is also where I found myself during my journey to Poison Ivy.

* * *

I stepped toward my usual portal tree but felt myself beckoned to the left, down a poison-filled path, instead. In real life, there really is a frightening amount of poison ivy on the overgrown paths of the abandoned camp. In my journey, the vines started to glow a bright green. I knew where the plants wanted me to go.

A few minutes' walk down that path was the old chapel. It was a Lutheran camp, and the chapel was a stone altar and cross between two rows of pine trees parallel to the river. As a child (I started going to camp when I was six), sitting on those soft pine needles as the sun was setting, I first found God, the Great Spirit, Magic, or whatever you want to call it. It wasn't in anything that anyone was saying. It might have been the singing. But it was mostly the feeling of being still and reverent in such a beautiful outdoor space, the scent of pine and the musty smell of the Delaware feeding my brain with love and the mystery of the universe. That is where Poison Ivy was telling me to go.

I sat down in the middle of the chapel with my back toward the altar. I watched as Poison Ivy started glowing and growing, creating a beautiful and terrifying leafy cathedral cage around me. I felt it closing in and was scared. I could feel all the fear causing my neck to tense up tightly.

"You carry fear in your neck," she said in a burbling-brook voice. "You don't need to carry it anymore." I released the fear in my neck and felt a glorious physical healing like I'd never felt before. I noticed that the glowing Ivy was not moving any closer to me.

"We create boundaries," she continued. "And boundaries make us feel safe." That is what my therapist keeps telling me too.

Suddenly my fingers started making funny movements, like they were wiggling and crawling. My arms began to rise as my fingers wiggled and I realized I had shape-shifted into a poison ivy vine climbing a tree. (For those of you who aren't familiar with poison ivy, her vines are "fuzzy," covered in hairy roots that attach to whatever she climbs.) I felt a loving connection to the tree I was climbing. "We protect what we love. We create boundaries to protect nature. We are the guardians of the forest." I could see it and feel it. The glowing green Ivy was creating a beautiful lacy chapel of plants — on the ground, in the trees.

"But what about us humans?" I inquired. "And the poison ivy on my daughter's steps?"

"Touching humans is not my love language," she joked. "And a little bit of fear is healthy. Fear can keep you safe. Those girls need to feel more fear. Setting boundaries also keeps you safe."

At this point, my thoughts were along the lines of OK, I've gotten enough here for my book. I'm done. But the drum was still drumming. What else was I meant to learn? What was next?

She appeared. Glowing green like the plants, but now in human form. She was perhaps thirty feet tall. A beautiful woman with a green smiling face and sparkling green eyes. Her head had a crown of poison ivy leaves. All of her was glowing green — her face, her hair, her flowing dress. She was a goddess.

"We teach humans to pay attention," she said. "To be present. Watch where you are going. Pay attention."

My hands started making another weird movement — almost like praying but also a rhythmic back-and-forth deep massaging. A Moravian prayer, which I can never remember in real life, kept repeating in my head:

> *Be present at our table, Lord;*
> *be here and everywhere adored,*
> *from your all-bounteous hand our food*
> *may we receive with gratitude.*

I held out my hands, and Poison Ivy started to pour down into them a shower of green sparkly energy. The whole forest was sparkling green and white.

"You don't need to carry your fear anymore. You are safe," she said, her hands sweeping the fear from my whole body.

I cried.

The journey ended at exactly that moment.

* * *

A twenty-minute journey. That's all it took to go from fearing to feeling a profound love and gratitude for the beautiful goddess Poison Ivy.

On the East Coast, the botanical name of poison ivy is *Toxicodendron radicans*. It's actually not an ivy at all but a member of the cashew and pistachio family. But DO NOT EAT IT. The ingredient that causes the rash is called urushiol. Don't eat that either. One of the messages I felt Poison Ivy was telling me was that we don't need to ingest things to find healing from them. The energy of the plant itself — the spirit of the plant — has the power to heal if we connect with it energetically.

Lest you think my journey was overly influenced by the DC Comics character Poison Ivy, I can assure you I have never seen her in any movies or comic books. I didn't know she existed as a narrative character until I was browsing poison ivy topics online after my journey. In fact, even though Ivy was clearly a she in my journey, the plant has both male and female reproductive parts.

But it's classic that humans would turn Poison Ivy into an evil villain just because she gives you a rash while protecting and feeding her forest family.

Thinking about stories of good versus evil and superhero movies reminds me of a model of human interaction called the Karpman drama triangle, first described by psychiatrist Stephen Karpman in the 1960s. In the Karpman model, all dramas or conflicts have three roles: the victim, the perpetrator, and the rescuer or hero. We each take on one of these roles whenever drama develops. Sometimes we play the victim, other times the hero. Humans involved in the environmental movement tend to see themselves as the rescuers who need to "save the planet," or save nature. But if there is anything these interesting times we live in have taught me, it's that there are no purely good guys or bad guys. *Everyone is flawed* (yes, even me). *Every* country has a dark history. *Every* culture has committed crimes against humanity. All humans are the perpetrators and victims. And often when we try to be the hero we just mess things up even more. It's much more likely that nature will save us, even though we probably don't deserve it, than that we will save the planet.

The shamanistic perspective is that we must *transcend* the drama triangle and rise above it, creating new types of stories and relationships that are about collaboration and partnership . . . healing the trauma from drama. Drama can act like a drug on our nervous systems. But ultimately, drama is an addiction that prevents us from living fully and joyfully. Let it go. You don't need it anymore. We can start now to live more consciously and peacefully if we decide to. But it starts inside each one of us by choosing a new story.

Here is a new story to tell: Poison Ivy is the protector of the woods. And if you *pay attention*, she won't hurt you. You are safe.

Thank you, Poison Ivy.

Snake

*There is no sin in pleasure if
it's partnered with love.*

JOURNEY: DECEMBER 6, 2021

S nakes do not annoy me. Nor do they scare me. I don't hang
out with them, but I respect them. I also understand that many
people are afraid of snakes. Some people equate them with evil or
sin. Others associate snakes with sexuality and wisdom. And who can
forget the snake in the Garden of Eden who tempted Eve into eating
the forbidden fruit?

I am fortunate that the only snakes I encounter in my garden are
nonpoisonous garter snakes — slim, brownish green, shy, squirmy,
and reclusive. We meet only occasionally, usually when I am about
to step on a stone or am walking in the warm grass, and we are both
surprised and then go on our separate ways.

I wasn't originally planning on journeying to meet snakes precisely
because they *don't* bother me. But I am fascinated by the Rainbow
Serpent, considered the creator of the universe by Aboriginal people
of Australia. There are as many stories of the origin of the Rainbow
Serpent as there are Aboriginal tribes or "mobs," but all of them are
associated with water — as is the rainbow itself — and creation. I was
curious what Snake would have to say.

* * *

Before I even reached my tree opening, I was slithering on the ground. I slipped into the darkness inside the tree. I found myself in a den of snakes, and it was quite comfortable and cozy. We were all snuggled up together and I felt happy. Loved. My tongue darted in and out, which helped me to seek out a snake I could speak to. I didn't sense the snake's gender clearly. Perhaps Snake was more female than male, but I decided it felt most right to call it "they."

They led me to a room where a teeny-tiny fire was surrounded by a circle of teeny-tiny baby snakes. It reminded me of a Far Side cartoon, although Snake was not wearing glasses or a hat or wig, and I had to smile. They started to speak and teach . . .

"Once, long ago, Earth was just rock, volcanoes, and water. Then plants and fungi started to grow. They created the soil. Soil is life, and what makes life possible here. We snakes evolved from worms, and were — and still are — both land and sea creatures. The earthly desires of sex, hunger, and pleasure led us to mate and create new forms of life. Soon, all sorts of animals were created. Everything mated with everything. That's how evolution happened and how humans were eventually created.

"Snakes represent the earthly pleasures of sex, eating, and desire, and we are the teachers, the holders of knowledge and wisdom. For many thousands of years, we were partners with humans, and with some we still are. But as human man tried to dominate — the way male animals do — he felt threatened by our power and knowledge. He did not understand it. He feared it and demonized it. We became the symbol of earthly pleasures as *evil*, because we bring knowledge and understanding, pleasure and love."

In my mind's eye I saw a spiral moving upward — a *rainbow* spiral — an image I had never seen before but seemed obvious once I did see it. "We are all evolving and traveling upward on the spiral of life. We are the Great Spirit's earthly pleasures, and it is through those pleasures that the Great Spirit evolves as well. What men don't yet understand is that domination does not bring love. Our job here as a whole planet is to evolve toward love. Earthly pleasures are the

Great Spirit's way for us to do that, but we must learn to do so with love and caring for others.

"Snakes also create the energy lines of Earth so that our traveling and tunneling connect all of us." At this point I saw snakes creating tunnels and energy lines all across Earth — above- and belowground.

"Many ancient cultures revered the snake by different names — because we came first and are the energy of creation, the keepers of wisdom and story. Humans must learn that domination does not bring love. And love is the only thing that satisfies our hunger. There is no sin in pleasure if it is partnered with love."

As Snake finished speaking, all the baby snakes slithered over and piled into my lap. It was adorable.

"Tell the stories," Snake said.

The journey ended.

✴ ✴ ✴

Whew. Sex. Food. Pleasure. The earthly pleasures — which not only delight us but keep us alive and moving forward and upward on the rainbow spiral of evolution. Don't we all want those things? Good food, great sex, and that wonderful feeling of loving and being loved?

When we look at our lives through the short-term window of our singular moment on Earth, the ability to dominate and control feels more important than it really is. But if you zoom out and take the longer view, it's easier to see that domination and control are kind of impossible. Ridiculous even.

Take sex, for example. We are a complicated species, especially when it comes to sex. The news is filled with sex scandals, accusations of pedophilia and sex trafficking, and reports of horrific rapes of children and women. Humankind has yet to develop a healthy attitude about sexuality — for many reasons. And yet sex is the true source of creation. Sure, humans can now create babies through in vitro fertilization and using surrogates to gestate and give birth. There is still a sperm. An egg. A uterus. But even after a woman is too old to have children, she can still experience sexual pleasure and orgasms. So clearly sex is not *only* about procreation.

As with religion, we may never discover one universal truth about sex and whether there is an absolute right or wrong. (In nature, though, sex is essential. Without sex, *everything* stops.) I am not going to go into details of the history of religion and sexuality, although I could. Nor the political insanity around controlling sexuality, although I could. I am going to stay focused on snakes.

How do snakes have sex? Interestingly, a female snake has quite a bit of control in the situation. First, when she is in the mood, she leaves a trail of pheromones to attract a mate. She can mate with multiple males, but *she gets to decide when and if to have babies*. She can store sperm for up to five years. (And snake copulation can take an hour or a whole day. Way to go, snakes!) With more than three thousand known species of snakes on Earth, there is a lot of diversity in the details of their sexual behavior. Maybe this is a reason why snakes are so threatening to so many people?

Snakes have been revered and worshipped in many cultures around the world. But the Abrahamic traditions — Judaism, Christianity, and Islam — revile the snake and associate it with the devil. The Virgin Mary, for example, is often pictured with her foot stepping on the head of a snake. I suspect such images were intended to "demonize" those prior religions that revered the snake, but I'm more interested in exploring new ways of thinking about how things *could* be — and looking to nature to figure out what is true. And for that, let's consider a snake's technique for renewal and growth: shedding its skin.

Snakes can shed their skins many times a year, especially when young. As they get older it happens less frequently. For the snake, shedding is an essential way to grow and regenerate, and it also helps them remove parasites. For humans, shedding one's skin is a powerful metaphor about change and outgrowing old mindsets or paradigms and making room for something new.

We are in need of a new paradigm, and perhaps we are already in the midst of a shift to a new view of the universe and our place in it. We are shifting from a culture of human domination over nature (and each other) to one in which we recognize nature as our essential partner. There are still many remnants of the old paradigm. But thanks to social media, which allows us to watch human activities unfolding anywhere on Earth in real time, we now can understand our human hubris as a pathetic attempt to destroy, steal, and control that which can never be fully controlled.

Demonizing things does not make them go away. Instead, we might want to shift our perspective and accept that diversity is an essential part of nature and embrace all our messy diversity with love. After all, as Snake said, "Domination does not bring love. And love is the only thing that satisfies our hunger." That is a big message for all of us. "Our job here as a whole planet is to move toward love." I think back to the rainbow spiral I saw during the journey. The spiral shape is found around the world in ancient art. It's also ubiquitously found in nature, from the curling shape of a seashell to the coil of a snake's body. Zooming out again, spirals are also found in cloud formations — a hurricane is a giant spiral — and galaxies, many of which are spiral shaped. In fact, while the conventional perception is that planet Earth simply rotates around the sun, the planet — and all of us on it — are actually moving forward through space in a spiral toward some unknown destination. At a very fast speed, I might add. Time is actually a spiral. A rainbow spiral?! Yes. A rainbow spiral.

Where will our spiral path take us? That is the mystery. Hopefully toward love. Love is the only moral code we really need to guide us. Snake tells us if we can marry our earthly pleasures with love, then there is no sin in them. That's the dream we need to dream.

"Dreamtime" is a term that Aboriginals use to describe their view of the universe. Some interpret it to mean the time of creation — the time of the Rainbow Serpent. Others interpret it as a mythical place where the world is continually created. One translation that helps me understand it better is *Everywhen*. According to the Aboriginal people, all time is happening all at once, and we are in a constant act of creation and dreaming the future into being —

and that understanding has the power to change our perspective of the past *and* the future.

My view is that the Rainbow Serpent is still with us, still creating the world and dreaming it into being. And we are all part of the rainbow. Sure, if you look closely, things are a horrible mess at most times, but if you zoom out, we are snaking our way forward and upward on the spiral of life. All part of the same journey, moving toward love in partnership with nature, who will not hesitate to remind us that nature is the only one who is really in control. In the meantime, there is no sin in pleasure if it's partnered with love.

Thank you, Snake.

Grass

Nothing can grow without love.

JOURNEY: DECEMBER 13, 2021

Grass. Americans are obsessed with their grass.

I am not anti-grass. A nicely cut lawn is perfect for playing, walking barefoot, and getting from one place to the next. Some people I know and respect have love affairs with their lawn and even their lawn mowers and deeply enjoy the ritual of mowing. But some have taken lawn worship to the extreme. Conservatively speaking, lawn takes up more space in America than the top three agricultural crops combined. (That's corn, wheat, and fruit trees.) Lawns also use a *lot* of water. Think about it for a minute . . . We devote precious water to a plant that we then mow, usually with gasoline-powered mowers, sometimes more than once a week. Not to mention all the toxic chemicals that people put on their lawns to get rid of "weeds," especially the dreaded dandelion (her chapter is coming up next). What a waste. I call it the tyranny of tidiness.

Grass is not bad. In fact, I would vote for grass being good. Grass is a plant. It has roots. It transforms sunlight and carbon dioxide into energy and oxygen and stores the carbon in the ground . . . if it's not sprayed with toxic herbicides, fungicides, and insecticides. Also, grass smells nice.

I have an acute sensitivity to toxic chemicals, whether they are agricultural chemicals, cleaning products, "beauty" products, or fake fragrances. I can't wear makeup or use most beauty products because they make my eyes burn and itch. Chemical fumes make the skin between my upper lip and nose tingle painfully and turn blotchy red. I don't go anywhere near the farm and garden chemical aisles in stores for this reason. Grass sets off my built-in toxin detector all the time. When I drive through a suburban neighborhood or past a golf course, I can *feel* the lawn chemicals — on my face!

When my former husband, Lou, and I first divorced, we took turns living at our home with Lucia. During my times away, when I wasn't traveling for work, I stayed at a cabin in the woods owned by the local conservancy. The previous tenant (OK, my mother) had relentlessly mowed the area around the cabin. I had no lawn mower, and no strong conviction to mow. Instead I let the grass grow. And that's when the magic happened. The cabin started to feel like a mythical European fairy-tale cottage. Rather than becoming unruly, the grass took on the look of a woman with long hair blowing in the breeze. Abundant wildlife became even more prolific. Watching the fireflies blinking by the thousands in the long grass at dusk was a magical dream come true. I felt like I had healed the land. I also healed myself.

But who and what is grass, really? I wanted to learn more.

Did you know that the grasses we grow in our lawns are only a tiny fraction of the twelve thousand species of grasses found on planet Earth? The Poaceae (grass) family includes all the prairie grasses, grazing grasses for animals (which can be turned into hay once harvested), rice, wheat, and even corn and bamboo. Suffice it to say, we could not survive without grass. Animals eat grass. We eat grass. We eat the animals that eat grass. We don't exist without grass. Grass is *good*. Very good.

I doubted that I could handle encountering twelve thousand species on a single journey, so I decided that I would talk with basic lawn grass. That's still dozens of species, many of which are also considered weeds, such as crabgrass or Bermuda grass. But I was also eager to speak with and better understand an ornamental grass called sea oats, because it *really* annoyed the hell out of me.

When I first landscaped my yard, I wanted to include native plants. I consulted a landscaper I trusted and had worked with for decades. I had a vision of a whole area of native grasses that would be easy to maintain and would wave gracefully in the winds. He recommended sea oats.

It's a pretty grass, although the seed heads now remind me of the brown marmorated stink bug (I planted the grass before that Asian species came to America). But within two years the sea oats was Out. Of. Control. Not only had it spread everywhere, but it was incredibly hard to dig up and get rid of. It has a knack for seeding itself in extremely difficult places — between rocks and nestled in the roots of my roses. Thus began Maria's Glorious Five-Year Plan to eradicate sea oats from my garden. It has not gone well. It's seventeen years later and sea oats are still everywhere.

Last year I went to a local native plant nursery to find some other grasses to plant in a different area. I asked the sales guy what to plant.

"Anything but sea oats," I said. "It's so invasive."

"It's not invasive because it's native!" he protested.

"What is it then?" I challenged him.

"It's *aggressive*," he declared smugly.

Oh, so that's the difference, I thought to myself as I rolled my eyes behind his back.

I decided to journey on a Monday morning. It was in December 2021, right after a giant tornado had destroyed areas in Kentucky and a few other states. So extreme weather was also on my mind.

I entered the journey by sliding down a mud chute. I landed in the grass. As I lay there, the grass consumed me and turned me into soil. Then I ceased to exist as a separate entity and felt completely absorbed into the earth. "We are hungry," the grasses said. I went down underground and could hear a global network of communication happening in the roots. Meanwhile my eyes were burning painfully. I kept needing to rub them.

"We are the network of the world. And we communicate with the weather to act on our behalf, asking it to help us." As Grass spoke, I could feel their hatred of humans for poisoning them. "Yes, plants can feel hate. We hate being poisoned. Hate it. And the weather feels our rage — hurricanes, tornadoes, droughts, floods, fires. When there is no love, there are no plants, no grasses. Without love, we recede, and all that's left is the desert and man's hubris." I saw an image of a desert, completely devoid of life.

"We can't grow without love. Nothing can grow without love." I saw tornadoes, droughts, and destruction clearing the way so that plants could return to those landscapes without poisons. I understood that man himself can be like a toxic chemical if he overhunts and overgrazes, if he lives without love for the land, nature, and women. If he continues to spread poisons everywhere.

"What's up with men and their lawns?" I asked Grass.

I saw an image of a woman being forced to dance until she collapsed from exhaustion and died. There was no joy in her dancing, just the frenzied panic of someone being forced to dance for someone else's desires until there was nothing left of her.

"That's not love, that's pornography!" Grass said. "Men look at perfect grass and weed-free fields the same way they look at porn — like it's some unblemished fantasy person they can do whatever they want to with, regardless of their desires and wishes. That's not love, it's porn! Men need to be taught how to love."

"How can we teach men to love?" I asked.

"It starts with being raised with love and letting him feel safe to feel all his feelings. Letting his heart crack open so the grass can grow free. But then it's about him learning to respect women and how to accept imperfection — in himself and others."

I felt the time running out on my journey. So I quickly asked: "What about sea oats?"

"Get a grip, Maria. Relax. We are busy establishing the connections and growing where other stuff won't."

Grass continued, "All the grasses need love to grow. And yes, Earth is a woman, but the plants are male." Then the journey ended.

✳ ✳ ✳

How to unpack this bizarre conversation with Grass? Eat breakfast, first. Then, let's go line by line. Blade by blade.

Blade 1: The grasses told me, "We are hungry." And it's true that grasses are a hungry species. This plant hunger is why chemical companies have put so much effort into coming up with synthetic chemicals to fertilize and manage grass crops. Yes, the same chemicals that are incredibly toxic to soil microbes and humans. The Rodale Institute Farming Systems Trial has shown that *animal manures* are the best fertilizer for corn (a grass), wheat (also a grass), and other grass crops, including pasture (more grass). Anyone who has had a lawn and a dog knows this is true — where the dog poops, a patch of greener grass grows. Same with animal poop in a pasture. This is why we need animals as part of our farming and feeding systems. People don't like talking about poop, but it's an essential resource that is free. Put the animals back on the grass and the poop on the crops and you don't need synthetic fertilizers. Unfortunately, though, human poop has become toxic due to our overuse of pharmaceuticals, chemicals, heavy metals, cleaning supplies, and all the other stuff people flush down the toilet that they shouldn't. This is why applying human sludge as fertilizer is rightfully not allowed on organic farm fields. (Do you know what else is a natural, super effective fertilizer for grass? Urine. Animal *and* human urine.) In addition to being hungry, grasses are very thirsty. I was happy to learn that the state of Nevada has banned "nonfunctional" grass in Las Vegas due to the extreme drought facing the region. (Although the ban exempts golf courses, homes, and parks and doesn't take effect until 2026, which may be too late.)

Blade 2: Grass said that grasses are the communication network of the world and they communicate with the weather. That's interesting, and to me, it makes sense. There is some scientific basis to the concept that plants and the weather are connected and communicate with one another. There is evidence that trees and their roots communicate with one other. Why not grass? When trees are planted where there were none before, they lower the temperature, increase the amount of rain, and generally create a more pleasant place for humans and nature to thrive. This connection between grass, plants, and weather seems like an interesting area for more research to be done. Perhaps Las Vegas should plant more trees.

Blade 3: "When there is no love, there are no plants." Looking back to the early history of life on Earth, there were plants before there were humans, so animal and plant love must count in this equation of no love = no plants. That's interesting. If plants can feel hate, they can also feel love. There are amazing research studies demonstrating that plants emit screams when they are harvested and can warn each other when danger (insects or diseases) is nearby. If *everything* has feelings — plants, animals, and humans — yet we *all* need to eat, then it seems to me to best way to proceed is to harvest and eat whatever we decide to eat with reverence, kindness, and deep gratitude.

Blade 4: "That's not love, that's pornography," said Grass, referring to the desire for a perfect lawn and to the woman forced to dance until death. I am not anti-porn. Porn can be a useful tool for learning and release. (But not child pornography, which is always horrifying and inexcusable.) Porn also can be an addiction. Equating the use of lawn chemicals and crop chemicals to pornography makes a lot of sense to me. Porn is not usually an expression of love. Nor is a perfect lawn or weed-free farm field an expression of love. It's an act of voyeurism rather than connection. Treating grass with chemicals to force the lawn to grow in unnatural ways is also asking grass to "perform" without providing real food, or allowing the plants to rest when needed, or considering what their desires might be.

While porn is usually consumed in secret, a perfect lawn or weed-free field is an outer expression of an inner fantasy of a person or world without blemishes, without independence, without creative expression. There are a lot of people — women especially — who spend an inordinate amount of time trying to look "perfect" to please others, usually men. But if there is no real love involved, for self or others, it's an empty fantasy, devoid of real pleasure. Women can suffer from the tyranny of tidiness too.

Now imagine soldiers marching in lockstep, like armies in historical videos or present-day videos from North Korea or Russia — military porn, if you will. (If an Iowa cornfield could march, it would look like an army marching.) It looks pleasing to the eye and "perfect," but who are those individuals? What are their desires, and are they happy? Obedience and order do not equal love and happiness.

Real love considers the needs of the object of our love. Real love is interested in *mutual* pleasure. Real love is willing to embrace the messy imperfection of another person or place.

I believe that the American obsession with guns is also a form of socially acceptable porn. I once received an articulate but anonymous letter from a reader of *Men's Health* suggesting this very thing. (I asked the editor to publish the letter, but he refused. Too controversial.) When people are forced to bury their sexual desires underground, those desires can burst forth as a desire to control others and act out fantasies. They may act on their fantasies in secret, or they may act them out publicly in more socially acceptable ways, like playing with guns, for instance. Or tragically, they may act out in socially unacceptable ways, like domestic violence and mass shootings. Sexual desire is a powerful natural force that cannot be ignored and expected to go away. It will always be channeled somewhere. Like water flowing downhill, it can't be stopped. The more we can let our desires flow into loving consensual relationships, the more productive and happy (and safe) everyone will be. That's why it's so important to learn to talk openly about sexuality and teach people how to express it in positive ways.

In the long run, sex without love will always feel empty. And what everyone really craves (even if they are afraid to admit it or don't realize it) is the love and intimacy that comes from a healthy view of sexuality and a freedom to express it with another person — with love — even if it's messy and kinky. Because truly great sex is often messy and a bit kinky. And that's nothing to be ashamed of.

Blade 5: Studies supported by the National Institute of Justice show that most domestic violence and school shootings are perpetrated by people who experienced severe trauma and or a lack of love in childhood, whether from sexual abuse, parental abuse, or social bullying (or all three). We all need love to grow and be healthy. In fact, "failure to thrive" is a recognized medical term for children who don't grow (physically, mentally, and emotionally), often (but not always) because they are denied love and care — specifically loving touch and emotional caring. *Nothing can grow without love.* To address the horrific gun violence in the United States (which is unparalleled in any other country), we must get to the root of the issue and have the true manly and womanly courage to protect each other and our children from harm. That's what love does.

Blade 6: "Earth is a woman, but the plants are male," Grass had told me. Now I get it. Life on Earth must be a partnership. One without the other is nothing. One with the other is everything. That is our work we need to do together — as men and women, as humans with nature. (That doesn't mean that everything must be male/female heterosexual. As I have learned from researching this book, there is so much sexual diversity among all species that diversity isn't just normal but perhaps it's required for some reason we have yet to discover.) It dawns on me that when the beings I speak with say their work is about keeping nature in balance, they might also be asking us to do the same in terms of creating true partnership between the sexes. Among all the sexes and variations of sexes.

I am humbled by these journeys.

And scolded, too, as Grass told me to "get a grip" about sea oats. OK, I will relax. Although "get a grip" sounds an awful lot like weeding to me.

Lest you think I am naive to believe a world based on love can work when it comes to restoring nature, please read *The Boy Who*

Grew a Forest, by Sophia Gholz. It's not a long book — it's written for children, but it's a story that everyone should read. (Or you can look up the story online and see the boy, who is now a man, tell the story himself.) It is the true story of Jadav Payeng, who completely healed nearly 1,400 acres of barren land in India by planting bamboo (which is a grass) and nurturing and loving the plants that followed.

Nothing can grow without love.

People, we have work to do. But together we can do it.

Thank you, Grass.

Dandelion

Just be happy.

I usually let several days go by between journeys because I like to ponder each one and give it space to settle in. But there it was, only two days after my Grass journey, when I awoke at 5:30 a.m. with a strong urge to speak with Dandelion. After all, how can you talk to grasses without talking to dandelions? They would not leave me alone until I came to visit.

While I drank my morning coffee and watched the sunrise, I tried to predict what they would say or how the journey would unfold. As usual, I was completely wrong.

✳ ✳ ✳

I entered into the tree, and it was dark. "Dandelion, dandelion, where are you?" I called out.

"Here we are!" they said joyfully. I was sitting cross-legged in the grass, surrounded by many yellow dandelion flowers that seemed happy to see me. They were dancing gently on their stems. Suddenly I felt a root coming out of my perineum and into the soil below. It was a bit alarming. I resisted at first, but then remembered all would be well, and I just let it happen. I began sprouting leaves, too, and I/she started to sing in a happy little voice: "Put down roots, nourish with our leaves, rise and shine, shine, shine. Then let our seeds fly, fly, fly into the sky like ballerinas dancing on the wind. Just be happy. Just be happy." I watched myself unfold into nourishing leaves. I felt myself rise up and bloom. And then I watched my fluffy seeds fly off and dance in the wind.

"What about all the people who try to kill you?" I asked.

"Oh those grumpy dumps, skalliopagulusses [or some such silly word]. They can't stop us! We will grow anywhere. They are only hurting themselves. Just be happy! Shine!"

I felt myself blowing air out of my mouth like I was blowing a seed head of a dandelion.

"Make a wish!" Dandelion said happily. I wished for what I always wish for — true love. But then I got sad, because it feels like my wish has never come true. "Just be happy!" she said again.

"What is true love, anyway?" I asked.

"It's when you learn to love everything as it is, not as you wish it were. When you love things as they are, then your heart fills up and fills up and then you shine!"

"But what about the grumpy dumps? How can I get them to stop killing you?" I asked.

She showed me a picture of a blackened, hardened heart that was cracked and broken. "Oh, no. You can't fix other people. That's their path to travel. All you can do is love them as they are. As it is. Just be happy."

Next I was back in the meadow, sitting cross-legged. Suddenly the dandelions started swirling around me in a spiral.

"And dance!" they shouted out as the drum called me back from the journey.

✳ ✳ ✳

Well! That was an interesting way to start my day.

Am I a grumpy dump? Sometimes. During the journey I had found myself wondering whether dandelions are perennials . . . so I looked it up. Suffice it to say there are so many good things about dandelions that it's hard to know where to start. Yes, they are perennials. And if you don't get all the root out when you pull a dandelion plant, it will regrow from just a tiny piece of root. Many kinds of dandelions are also asexual, which means they don't need pollen from another dandelion (or even their own flowers) to make seeds. All their seeds are clones of themselves.

Dandelions have been feeding and healing humans since forever. All parts of the plant are edible. The roots can be made into a caffeine-free coffee-like drink. The leaves are high in nutrients and, while bitter, are delicious when cooked and served properly. With a Pennsylvania Dutch mother, I grew up eating dandelion salad with hot bacon dressing in spring, when the leaves were young. She called dandelion "piss-a-bet," which means what it sounds like because the leaves are a diuretic. Last summer for the first time I fried up some dandelion flowers (dipped in batter), and they were surprisingly delicious as well. You can even make rubber from the plants' milky sap.

The Pilgrims brought dandelion seeds with them on the *Mayflower* because the plant was considered essential for nutrition and healing. Basically, any health problem you think you have can be helped by eating, drinking, or including dandelion in healing potions. But don't use dandelions that have been sprayed with chemicals (obviously).

As I think about Dandelion's message, I have to wonder: Could life really be as easy as "just be happy"? I'm starting to think that it might be. We can't fix other people; we can only fix ourselves. And fixing ourselves isn't so much about "work" as it is about learning to love, relax, and connect to others with kindness. "Trolls" exist because we respond to them. If we ignore them, they have no power. For all our sakes, resist the urge to insult or ridicule people. No more being snarky! So often people say hurtful things they think are clever or funny on social media. But the real effect of their words is to drive the knife deeper and deeper between people (on all sides of an issue). *Mean humor hurts.* And it's not really funny either. I first realized we were headed for trouble in this country when every person I talked to (mostly liberals) was getting all their news from comedians. Sure, it was all funny and interesting. But it was also deeply hurtful to many people. Hurt people tend to lash out. And here we are, America.

As the Covid pandemic stretched on (and *nothing* seemed funny anymore), many people started questioning what they truly want from their lives. I have heard lots of stories about people who decided to do what they love — what makes them happy — and the universe provided for them, and they found success and joy and abundance. I can relate to that. At times when I have felt deeply depressed, I realized that I had veered too far from what brings me happiness and joy in my life. Making life changes (and therapy) has always brought me back to joy.

There have been a few times when what I call The Darkness was very, very dark and frightening for me. In every situation, getting out of The Darkness involved making difficult decisions that required courage and heartbreak. I had to quit jobs. I had to get divorced. I had to decide that the family business should be sold. I had to get out of a relationship that was not giving me what I thought I needed most. In every case, while the separations were excruciating, life on the other side was brighter, better, and just plain happier.

Sometimes The Darkness was (and is) due to things outside of my control — politics, environmental concerns, or the ongoing horrors of humanity, like wars, violence toward women and children, the suffering of those who are hungry and sick. But in each case, once I asked myself what I could do and started to do it, I felt more empowered (and less sad). Small steps, like working at my local election polling place to get to know my neighbors better. (There is a big need for poll workers. You, too, can protect our democracy and the vote.) Everyday choices, like being kind to people even if I don't agree with their politics, refusing to use chemicals on my property, and supporting local businesses and people that are attempting to do good things. And sometimes taking on a big project, like writing this book to give nature the voice nature has been asking me for.

Today, my lawn is sprinkled with dandelions (and clover, which is a natural nitrogen provider). And instead of feeling annoyed, I feel happy.

Maybe just being happy is enough to get started. Figure out what makes you truly happy and start there.

May you find joy in the journey.

Thank you, Dandelion.

Weather

Morality must be innate if humans are to survive.

JOURNEY: JANUARY 1, 2022

As I write this, it's an extremely foggy New Year's Day 2022 in Pennsylvania. I don't mind the fog. (In fact, my teenage daughter says my clothing color palette is "foggy day.") There were record-breaking fires in Colorado yesterday — over a thousand homes burned to the ground. Record-breaking tornadoes in the South in December, and 2021 was generally a year of extreme floods, droughts, heat waves, and storms. Is this a tipping point where we begin to take climate change seriously? Doubtful.

Few things obsess humans — especially those of us who try to grow food — more than the weather. One of the reasons I have decided not to move somewhere else is that the weather in Pennsylvania is reasonably stable. And on those maps that show how climate change will affect the future, where I live now is in a prime spot — high enough to avoid flooding and far enough north to avoid the worst of heat and drought. Sure, I'd love to live on a beach. And maybe one day I will (without even moving). But for now, I'm just trying to figure it all out and stay calm.

My Trail of Books led me to read a book on shamanism called *The Re-Enchantment*, by Hank Wesselman. Wesselman mentioned a book called *Weather Shamanism*, by Nan Moss and David Corbin, which I realized I had bought years ago but never read. I knew right where it was on my library shelf. I read it, and some of what I learned from it made me really, really want to talk to Weather. I didn't know if it would be possible to do so. Even though the weather impacts us every day and we all talk about it all the time, it hadn't occurred to me that it could be a being I could speak with. But I wanted to better understand weather, not to control it or fix it. Not to make the fog disappear, but perhaps to clear the fog in my head about how we can have a better relationship with weather in general.

I decided to make Weather my first journey of the year as a way of honoring this being. And you have to understand . . . I am an obsessive New Year's resolution maker, planner, and list maker. But I made this journey before I even began my planning and lists. Perhaps that's because I knew that Weather has the power to upend any plans I might make anyway.

So, happy New Year, and away we went . . .

✳ ✳ ✳

I was in a desert, and a sandstorm lifted me up to a platform in space to look down on Earth. It was a little nerve-racking, but I felt a masculine presence next to me telling me to relax. *Relax.* As I looked down on Earth, he spoke.

"We make the waves. We cleanse Earth." He shrugged a little and continued, "We don't really care about people that much, we care about keeping Earth in harmony."

"How can we help?" I asked.

"Kindness. Gentleness. *Reciprocity.* We care about actions and behaviors. You don't need your outdated religious books. You don't need to make sacrifices to us or build altars or monuments. *Your morality must be innate* if humans are to survive. And money is meaningless, worthless. The only thing that matters is action and behavior. Respect. Kindness. Gentleness. Appreciation."

I watched the weather swirling and circulating on Earth. From my vantage point in space, no humans were visible there.

"We are love made visible," the voice of Weather continued. And suddenly I saw it and felt how weather encompassed all the facets of love. He continued, "We are the glorious love of a perfect sunny day. The painful love of a storm. We are the force that creates and destroys, and nothing can get between us and Earth."

I was quiet. But gently, I asked what I could do. "Dance, Maria. One dance a day. One song. That's all we ask of you."

Now I merged into a storm, and I felt his personal responsibility to *protect Earth* above all things.

"It's our job is to keep Earth alive, not yours. And that's exactly what we will do."

Thus spoke the Weather.

✶ ✶ ✶

I came out of the journey with a song in my head that I knew I had to dance to.

After mostly failing out of public school as a teenager, I went to boarding school for my junior and senior year, and there I discovered a radio station that saved my life. WSLE out of Peterborough, New Hampshire, was a folk station with gentle-voiced DJs who brought into my life songs that would stay with me forever. It was also in boarding school that I learned how to dance with abandon and without shame. I took a modern dance class. There were no judgmental family members around to shame and embarrass me, so I *danced.* I became known for my dancing. I even won an award for it.

But then I graduated and stopped dancing. A series of sad events made it hard for me to do any sort of dancing unless I was drunk. And then I stopped drinking. But music kept my heart alive, even if I was only dancing in my head or at super loud rock concerts (which are essential to my joy).

Fast-forward several decades. I was walking in total darkness on a cold autumn night next to the shaman Alberto Villoldo at Kripalu, a spiritual retreat center in Massachusetts. We were headed to the place where we were going to do a sacred fire ceremony. I'd only read about fire ceremonies before, so I was excited to experience one firsthand with Villoldo. He didn't know anything about me, and I sensed that he didn't really care to. But I knew who I was and how important this moment was to me, and so I walked right next to him in silence, ahead of all the other classmates.

He opened sacred space and lit the fire. And then he told us he would teach us the sacred fire song to sing together. He began . . .

"Witchi-tai-to . . ."

Wait, what?!?! I knew this song. This was one of my favorite songs by Brewer & Shipley, a folk rock duo. I listened to it on WSLE. I still have the vinyl album. This was another one of those moments that Lisa would call YCMTSU: "You can't make this stuff up." (Although she informed me that Villoldo actually sings "Nitchy-tai-tai," but it was the exact same melody.)

I sang along like I hadn't been able to sing along to anything in ages. I'm always uncomfortable chanting in other languages and singing in front of others. But this song, this chant, I knew. I loved. I understood. I sang.

The song was written by Jim Pepper, a member of the Kaw and Muscogee Creek tribes. The words were taught to him by his grandfather as a peyote ritual song. Although he no longer knew what the Kaw words meant, Pepper added words in English about the water spirit, which is what carries the peyote visions. I only did peyote once, in high school, not in a sacred setting, and I don't remember anything from that. But I do know the water spirit. That's my friend the Rainbow Serpent!

After my Weather journey, I danced a New Year's Day dance to "Witchi-tai-to," and I don't care if anyone thinks I'm weird (although fortunately I was alone in my house). I realized as I danced that I am making waves in the air just like the weather makes waves. Waves are life. And music is waves. What a wonderful gift we have in our hands and bodies.

Day four into my New Year's Weather dance resolution, and I felt like I'd been released from a cage I'd been housed in for forty years. I dislodged a memory of the aerobic exercise era, when I felt forced to make my free-form dancing joy a procedure of exercise with rigid steps I felt unable to follow. This is probably why *Strictly Ballroom* is my favorite movie ("a life lived in fear is a life half lived!"). There are as many kinds of dancing as there are weather. All are good. All have purpose. All bring joy.

Another important message from the Weather journey still reverberates in my mind: Your morality must be innate if humans are to survive. I had been talking recently with someone who works with many coders in the tech field. He is deeply worried because they tend to be amoral. Not immoral, which is purposefully bad. Not moral, which is purposefully good. But amoral, which means they don't care about whether something is good or bad. These are the people who are creating our worldview through their technical expertise — our social media, our algorithms, our artificial intelligence. He is deeply concerned because, let's face it, organized religion has lost its moral authority. Between the sex scandals, the corruption, the violence, the authoritarianism, the small-mindedness — religion no longer speaks to many of us, and when it does speak, it doesn't make sense to those of us who have read the original books. This doesn't mean religion is bad and shouldn't exist. Religion serves many purposes and is a personal choice that many make. But it has lost its power to provide moral guidance — if it ever really had it in the first place. What can take its place?

Kindness.

Gentleness.

Reciprocity.

Appreciation.

Love.

Truth be told, I had to look up the word *reciprocity* just to be sure I knew what it meant, as it's not a word I often use. I went to my old-school *Oxford English Dictionary*: "A mutual exchange of advantages or privileges as a basis for commercial relations between two countries." Or between two people or even two entities, I would add. There is also the reciprocity principle of psychology, wherein someone is much more likely to be kind and generous to you if they feel obligated because you have already done something nice and kind for them.

In her book *Braiding Sweetgrass*, Robin Wall Kimmerer writes this about reciprocity: "Each person, human or no, is bound to every other in a reciprocal relationship. Just as all beings have a duty to me, I have a duty to them. If an animal gives its life to feed me, I am in turn bound to support its life. If I receive a stream's gift of pure water, then I am responsible for returning a gift in kind. An integral part of a human's education is to know those duties and how to perform them."

I learned from Lisa that in the Quechua language of the Q'ero, the shamanic tradition of the high Andes, a fundamental tenet of their cosmology is *ayni*, or reciprocity. "If we are not 'in *ayni*' with all, then we are out of balance and susceptible to illness and unhappiness," she told me. (I did not know this before my journey.)

It's simple, really. And utterly beautiful. Kindness. Gentleness. Reciprocity. Appreciation. Love.

I think a lot of the purpose religion serves is to help us get our heads around what might happen after we die — whether it's heaven and hell, reincarnation, or nothingness. Amorality might stem from not caring one way or the other what happens. And the truth is none of us know for sure what happens — although there are many stories, documentaries, and legends that taunt us with the potential options.

I was raised as a Lutheran. But my mother and I both quit the Lutheran Church when I was fourteen after being subjected to a horrible, unacceptable, misogynistic sermon (clearly, the pastor had had a fight with someone that morning, and his rage was palpable). My father had been raised by his Jewish father, who no longer practiced Judaism, and his Catholic mother, who was no longer Catholic, so he was given a choice of what to be. He chose nothing. But later in his too-short life he became very spiritual, and so was my mother. All this left me a bit untethered and free to explore my own meaning, which I did with gusto.

Do God and the afterlife exist? I don't know and won't know until I die. But I have always been strategically minded and believe in hedging my bets, so early on, I made a deal with myself. I decided that I would live my life as if God does exist and the afterlife is real and there might be consequences for doing bad things. That way, I have my bases covered. If there is nothing after death, at least I've lived a good life and done as little harm as I could. And if there is life after death, then I will know I've done my best. It's a win-win situation. The key is that I have made my own decisions about what is bad and what is not bad. For example, I believe sex is good. Sacred even. Hurting others physically, emotionally, or sexually is bad. Very bad. Truth is good — even hard truths. Lies are bad. It's not that complicated, really. Most importantly, I believe that everyone should have the freedom to decide for themselves what to believe. That's the glorious fun and mystery of life!

The more I journey, and the more I read and understand the sha- manic world, the more I believe that death on Earth is not the death of life and consciousness. In *Cave and Cosmos*, a book that Dr. Michael Harner wrote at the end of a full lifetime of anthropological research, he stated that in all the journeys he witnessed, researched, and read about, there was absolutely no evidence of a "hell." If anything, he believed, as I do, that hell is here on Earth. However, there is plenty of evidence from his research that there are other realities beyond this life, and they are similar across cultures, continents, and religions. Therefore, it behooves us to act with love. It's selfish, and that's OK.

Innate morality is not dependent on some book or authority. It comes from deep within us, born into us from an evolutionary perspective, cultivated over generations through repeated behaviors and actions.

I believe in science. I also know that sometimes what is considered scientific truth turns out to be wrong. This is why personal experi- ence and validation are so important but also confusing. What works for one person might not work for all. Sometimes what is true is hard to ignore (and shouldn't be ignored). I have a daughter who is involved with climate science research and has had the chance to validate evidence of climate change firsthand. She has lived on a glacier in Alaska for weeks, taking core samples of the ice, witnessing the magnitude of the melting. She works for an organization that monitors the oceans and, through that work, has seen the changes up close and understands the potential dire consequences of the heating, acidification, and pollution of our oceans to all aspects of our lives. This is serious, people. And while science can warn us, science isn't going to save us. Taking action and changing our behaviors will save us.

What behaviors?

Kindness.

Gentleness.

Reciprocity.

Appreciation.

Love.

We can't stop the weather from changing. But we can stop doing the things that provoke it and make it worse. We can start right this minute to bring a new morality into our existence, one based not on strict rules or outdated texts, but on love. Kindness. Gentleness. Reciprocity. Appreciation. Love.

It's the innate morality that will enable us to survive through any kind of weather.

Thank you, Weather.

Mites

Delight in your work.

Sometimes I wonder whether what I have learned from my journeys is personal or universal. For example, if a thousand humans journeyed to talk to a certain plant or animal, would the messages from the beings be consistent? Could the messages be scientifically validated? Does it matter? Maybe. Would that stop me from exploring these realms? No. The personal, anecdotal singular is enough for me. Journeying brings me great pleasure, and I share my experiences with you as a friend to a friend.

I will not be surprised or intimidated if some people ridicule or dismiss shamanic journeying. Scientific discovery always starts with one person playing around with an idea, and that person often is called crazy at first. My grandfather dealt with decades of ridicule and disdain and barely made a cent off the now $62 billion organic food industry for which he planted the seed.

You could say the organic industry began in the kitchen at my grandparents' farm in Allentown. And it was my grandmother who did the cooking (and gardening). My grandparents experimented on themselves first, following their passion to figure out whether there was a real connection between soil and health. N of 1, it's called: a single individual who represents the experimental test group. My grandfather was searching for ways to improve his own health because he wasn't finding the answers from doctors. He read everything he could and decided to experiment with growing his own food without chemicals to see if it made a difference. It did. But it took science another seventy years to prove it — mainly due to corporate forces that worked to deny the evidence because so much money could be made selling agricultural chemicals to grow less-nourishing food and even more selling pharmaceuticals to fix up the health of poorly nourished consumers. I inherited my grandparents' stubbornness and independent-minded nature, and you are about to witness that in action . . .

For decades I "outsourced" my housework so that I could focus on my job. But after I sold the company I decided to do all my own house-cleaning. I enjoyed being responsible for everything in my home. If something wasn't done the way I wanted it, I had only myself to blame. One day I thoroughly cleaned the bathroom next to the kitchen. The following day I walked into that bathroom and spotted what looked like sawdust on the toilet tank lid. I looked up. Usually when I see sawdust, I will also see carpenter bees. Nope. No holes in the ceiling (which wasn't wood, anyway). I wiped away the sawdust.

The next day I went into the bathroom, and there was *more* sawdust. I looked more closely and realized that it wasn't sawdust at all, because it was *moving*. Teeny-tiny bugs were swarming all over the top of my toilet. GROSS! I took a picture and googled it, and off I went, down the rabbit hole of all sorts of scary things — blood-sucking bird mites, maybe? I got out vinegar and soap and even some alcohol and cleaned everything thoroughly. I checked all around to figure out where they might be coming from. I didn't see anything.

They came back. I poured boiling water on them. They kept coming back.

The more I read online, the more fearful I became. Bird mites can take over every room in your house, and they bite you! I start washing all the laundry in hot water and buying mite protectors for our bedding. I retrieved from the basement the old microscope that Elvin had given my kids. We took a scraping, magnified it, and took a picture. Ewww! It was definitely a mite of some sort. But the people who run the insect identifier app I paid good money for said, "Sorry, we don't do mites because they are not an insect." (Mites are in the arachnid family, which includes spiders and ticks.)

Every day the mites came back. I went to the hardware store and bought every natural miticide I could find. I tried essential oils. They still came back. I had a plastic bag next to the toilet filled with dead mites stuck to duct tape (my best mite-catching tool). I was ruthless. The bag got bigger and bigger. My mite anxiety was a constant torment. I could not stop until I had solved the mystery.

Finally, and for the first time in my life, I called an exterminator. The exterminator showed up and said "Yup, that's a mite. But I have to ask my boss to identify what kind." He also told me the only choice would be to spray the whole house, like spraying to kill bedbugs. "The spray is pretty toxic," he grimaced.

Please understand . . . I have a clean house. I am fastidious. I was starting to feel like this was some sort of test. I was mentally mocking myself. "Sure, Maria, you always talk about finding the root cause until it's in your own house, and then you are willing to kill and destroy!"

The next day the exterminator called back.

"My boss says they are mold mites. You can't spray for them — you have to find the root cause of the mold."

Gad damn! I felt so relieved, because mold mites don't suck your blood. But then I became even more curious. Where was the mold coming from? I didn't see any mold. So, we got aggressive. Elvin and I removed the toilet and took it outside. I checked the bathroom floor and the pipes into and out of the toilet. A tiny bit of mold, but not much. No sign of mites. I went outside with a hose and sponge to examine the toilet more closely.

At first I couldn't see anything unusual. Then I turned the toilet over so I could look at the back — the part that faces the wall. "That's weird," I thought. "Why is there dog kibble back here?" I rinsed out the kibble. And then I saw it: a hole into the interior of the toilet that was filled . . . FILLED . . . with old dog food, mold, webs of mite habitat, and other brown mystery things that I didn't care to examine more closely. I stuck the hose into the hole and turned on the water. For a good five minutes, brown stuff came out. Not human brown stuff, but other brown stuff.

Suddenly it all made sense. I had discovered the root cause of the mites. But how did the dog kibble get there in the first place?

I did have a dog at the time. She was my daughter Maya's dog, Lady Miss Penny, an adorable Shiba Inu who was very elderly, deaf, blind, and in a state of mental decline (poor thing). She was spending her retirement in my country home since Maya's city living was too stressful for her. I practiced unconditional love with Penny, and it was not always easy. All spring, summer, and fall I would leave the kitchen door open so that she could come and go as she pleased. Her bowl of dry dog food was right by the door, which was right next to the bathroom with the kibble-filled toilet. Apparently, generations of mice or some other rodent had been using my toilet as their winter pantry. Mold set in. And the mites were just doing their job to devour the mold. Mystery solved. Root cause uncovered.

If I had given in to my original panic and sprayed toxic chemicals to kill the mites, the only ones that would have been harmed in the long term would have been me, my family, and my pets. As exhausting and frustrating as the discovery process had been, I could now get back to living my life in peace and changing my own behavior to prevent more mite problems (not leaving dog food out overnight, for example). Everything in nature serves a purpose, and when we mess with it, we mess with ourselves. We can't save every bug or mite. But we can seek to understand, rather than assuming that a drastic response is the best way to go.

Those mites had been trying to tell me something. They were my teacher. Which doesn't mean they were welcome in my house (or that I liked them). As for the mice, well . . . a house is not a home without a mouse. That's country living. Elvin, though, was convinced it wasn't a mouse.

"Mice don't store food," he said. But I didn't want to think about some other marauder in my house, so I simply put the dog food away every night and went on with my life.

That was not the end of the story.

Lady Miss Penny passed away when winter came. It was sad, but she had lived a great life of sixteen years, and she was deeply loved.

I installed a new toilet in that bathroom, one that had no hidden holes to stuff food into. (The lady at the plumbing store thought I was nuts in my obsession about holes — toilets without holes in the back are not easy to find.) Because the dog had died, I was no longer putting food out (although Pumpkin the cat was still being fed in the basement). There was, most thankfully, no sign of mites. But then, while cleaning one day, I lifted the pillows of my *brand-new* couch and there, in the corner, was a pile of pet kibble. Holy Mother of Mystery! What the heck?!

Again, I took an unprecedented action. I bought rat traps — the kind that look like classic mouse traps, but bigger and more deadly. I placed a few around the corner of the couch and baited them with cat kibble . . . (Well, Elvin baited them. I'm terrified of setting traps. I don't mind removing traps with dead things in them, but the potential snap while setting them scares the bejeezus out of me.)

I'm sorry in advance for what I am about to tell you. If you are at all squeamish you might want to skip the next few paragraphs . . .

For a few days I checked the traps every morning, but nothing had happened. Then one day, in the next room, I found an empty rat trap. "Hmmm," I thought, "that's strange." I looked closely. Ah, the trap wasn't *quite* empty. There was a piece of fur, skin, and muscle still attached to the trap. Nothing else. No blood. No guts. No whiskers.

As a woman raised on Nancy Drew, after-school TV soap operas, and romance novels (the kind with mysteries included), I used my powers of deduction and pieced together what had happened: Based on the small piece of fur, which had a few white hairs among mostly brown, I determined it had once been a chipmunk. It had probably gotten caught in the trap without actually being killed and had dragged itself and the trap into the other room. The noise most likely alerted Pumpkin, who abandoned whatever else she was doing in the middle of the night and expertly expedited the death of said chipmunk and dined on whatever she could remove from the trap. Pumpkin never met a chipmunk she didn't want to devour. Mystery solved.

A year later, I found myself unexpectedly compelled to recount a much-condensed version of the chipmunk story to an eye specialist in Philadelphia.

My youngest daughter had been suffering from an autoimmune disorder of the eye called uveitis. It seemed to be chronic, and in 50 percent of cases there is no known cause. We had been to see this Philadelphia specialist three times, and he had dismissed many of my questions. Dare I say it, his responses felt a bit condescending. I am happy we found this doctor, and I'm very grateful to him. But I couldn't yet accept that there was no other option than keeping my daughter on steroids for the rest of her life (or worse; uveitis is the third leading cause of blindness). As a teenager, she was a bit resistant to my quest to find answers, and I respected her wishes. But I still wanted to find the root cause of her inflammatory response. The doctor, one of the best in the field, assured me that he had ruled out all the usual causes of uveitis with multiple levels of tests. He shrugged his shoulders, saying that we don't know what causes this problem and will probably never know.

Now, I knew this doctor's time was precious, but I launched into the story. I noticed that even the nurse turned away from her computer to listen. When I had finished, I leaned in and said: "Doctor, we are going to find that chipmunk."

(I left out the part about the cat eating it.)

Suffice it to say, my daughter was mortified. The doctor probably thought I was insane. But the nurse would most likely remember me next time . . . as the crazy chipmunk lady. I was OK with that. After all, I was an explorer and adventurer in search of the mystery. And I was a mom on a mission.

It's been two years since the mites appeared on my toilet, and now it's time to do a journey and see what I can find out about mites. When I journey, I often set my intention as a question: "What do I need to see that I am currently not seeing?" Let's "see" what I found out . . .

I still entered each journey a bit skeptical and nervous about what I might experience. This time, it was because I'd never tried to talk with an arachnid, and they did make me feel squeamish.

✷ ✷ ✷

I headed into my tree as usual but was immediately transported to the inside of my toilet, to the time when it was filled with mites . . . and mold. But instead of feeling disgusted and grossed out, I felt joy and a deep sense of productivity.

"We sing while we work!" the Mites giggled, "but you can't hear it because you're too big. We delight in our work!"

I felt their delight. But my eyes started itching terribly. The Mites told me that they don't see very well. I, however, could see a chipmunk with fat cheeks sticking his head in the hole and spitting out his load of dog kibble.

"By the way, those were our children you killed up on top of the toilet," the Mites said. "We were so happy and well fed that we had many children and there wasn't enough room for all of them, so we sent them out in search of new mold sites. No mold. No mites."

I apologized and thanked them and went in search of Chipmunk.

He was still hanging out in the room in my house where he had met his demise. He was a sassy little thing — fat belly, fat cheeks. He was reclining on the floor with his elbow out and his forearm holding up his cute little head. I think he might have had a piece of grass sticking out of his mouth. I asked him what he had been thinking, putting dog food in my toilet and couch.

"Hey, it's warm in here. Food was easy to get. What's not to love? Food, shelter, family — that's what it's all about. Delighting in your work to get food. If you don't find delight in your work, don't bother doing it."

I apologized for my cat eating him, but Chipmunk didn't seem too concerned. "We like to feel useful," he shrugged.

Then he took me into his den outside in my yard. In the summertime, I see chipmunks running into and out of the den's entrance all the time. But it is winter now in this journey, and there was a mom sleeping and snuggled up with a litter of babies. There was a dad too. "That's my son," Chipmunk said with pride. "He's a good lad."

I asked if he had any advice for humans, and he said, "No. We don't think about you that much. Cats, yes. Humans, no."

I thanked the Mites and Chipmunk, exactly as the drum called me back.

✳ ✳ ✳

Delight in your work! I was not expecting that.

I looked up what baby chipmunks are called. They are called "pups."

Then I tried to look up whether or not mites have a sense of sight, but all I could find was information about how to kill them. I did come across an article that said that black mold (which was the kind that was in the toilet) is known to cause eye inflammation in humans! So maybe black mold is a "chipmunk" that can cause uveitis — it's possible. (I've also considered the idea that I might be the chipmunk since I can be annoying, especially when I'm trying to get answers to things.)

I have since learned that we human beings each have over a million mites *living on our face*. Fortunately they are so small we can't see them. And apparently they do important work so we shouldn't try to kill them. Think about it . . . there are a million mites on your face right now delighting in their work. And singing!

I was not going to solve the mystery of my daughter's uveitis this day. And ultimately, finding her health solution is her own journey, just like it was my grandparents' journey. (Thankfully Lucia is off the steroids now.) I'm grateful to my grandparents for launching the modern organic movement and a huge natural health movement. I'm also grateful for doctors and the pharmaceuticals that eased Lucia's uveitis. In fact, without doctors and pharmaceuticals, none of my children would have been born and I would not have survived childbirth. What I understand fully now that I didn't when I first encountered the mite invasion in my bathroom is that finding the root cause of things can be hard, challenging work. But we can delight in it and laugh about it. Sing while doing it. Working together to find solutions to our health and environmental issues is a team effort. And rather than demonize each other, let's delight in our work *together*.

Thank you, Mites. And thank you, Chipmunk, too.

Tick

Relax.

Trepidation.

I feel trepidation about traveling to speak with ticks.

But I know I must. If this book is to be complete, I must venture to seek out the scariest thing in my garden.

Scary because they suck blood, of course, but even more because they spread Lyme disease. Is Lyme disease an engineered bioweapon? A few people have told me it is. And so, along with my fear, I feel curious to see what I will learn from these little bloodsuckers in the spider family.

OK, let's do this . . .

✶ ✶ ✶

I walked into the darkness of my tree and felt myself floating slowly downward. I landed in a grassy savannah next to my power animal. I noticed a tick on their neck, so I pulled it off and held it in my hand. It was engorged, and its tiny legs were squiggling. I asked to speak to it and lay back against the warm stomach of my furry friend. I started shrinking and shrinking until I was as small as a tick. Maybe I even was a tick.

"I am alone. I am lonely." I spoke Tick's thoughts aloud.

"I don't know my parents and will never know my children." I thought about how ticks are similar to octopuses in that way, never knowing their children (or parents) — and they both are eight-legged creatures too.

"We seek out warm-blooded beings (and cold-blooded sometimes too) to suckle and feed us. Just the way you like to snuggle up under warm blankets, we love to nestle up to warm skin to feed."

I asked what ticks' purpose was.

"Do we need a purpose? I guess we are food for birds. Opossums consider us a delicacy. When nature is in balance, we are just a tiny nuisance."

"Do you have a message for humans?" I asked.

"Relax. Stop trying to control everything. We are not going away and don't mean to cause disease and make you sick. We don't *kill* people. Just relax. And groom yourself and each other. It feels good and makes you pay attention to your body."

At this point I saw an image of a mother and child chimpanzee grooming each other. They were loving and caring.

"Examine your body. Aren't we amazing? *We are amazing.*"

At that point I found myself nestled back up against my spirit animal, and I hugged them so hard and told them I loved them just as the drum called me back.

✳ ✳ ✳

I confess I had a palpable sense of relief when that journey was over. See, that wasn't so bad. Ticks are lonely, just like us.

Just to be sure that ticks do live a lonely life, I checked into it. It's true. The male tick dies immediately after mating. The female tick dies immediately after laying eggs. They are a solitary and ancient species. Over a hundred million years old. Humans are still just a baby species in comparison. I think we are at the toddler stage, where, without strong parental guidance, we might well kill ourselves by doing stupid, preventable things.

My greatest fear when I moved to the woods was ticks. Especially the tiny ones. I cussed when I pulled them off my pets. I kept a jar of rubbing alcohol in my pantry for all the vagrant ticks I found crawling over me, my furniture, and sometimes my kids. Once or twice, I had to pull one off myself (usually after letting a pet sleep on my bed, a habit that was immediately quashed). For about six years I raised guinea hens, which eat ticks. But when people tell you guineas are loud, *believe them*. They are loud at all hours of the day and night.

When my youngest was three, we discovered, almost by accident, that she had Lyme disease. During her annual checkup, we noticed a strange rash. It was *not* a bull's-eye rash, which is all the "authorities" describe as the characteristic shape of Lyme disease rashes. It was an oddly shaped oval on her leg, with no sign of a tick bite in the middle of it. Fortunately, we had caught it in time, and she took antibiotics and had no lasting effects.

In my vigilance, I started a nightly habit of standing naked in front of a mirror and checking myself for ticks before I went to bed. They like warm, moist places like underarms, the groin area, and, for women, below the breasts. That's how I spotted the teeny-tiny tick that caused my own case of Lyme disease. My eye caught a tiny black spot in my pubic area. When I investigated closely, it fell off and I saw that it was a tiny deer tick, which are the worst for carrying diseases. I put it in a baggie and told myself that if any strange symptoms developed in the next few days, I would go to the doctor. Sure enough, two days later I got a flat, scaly, red rash, but on my shoulder, nowhere near the bite site. And it was not a bull's-eye. Nevertheless, I tested positive for Lyme disease and had to take antibiotics for twenty-eight days and stay out of the sun (hence the interruption of my previously described war on mugwort). I'm fine now. And it's true — people rarely, if ever, die of Lyme disease . . . thanks to the effectiveness of antibiotics. Although it does cause devastating illness in some people, especially if it isn't diagnosed and treated right away.

I think back to what I learned about the best way to reduce the spread of garlic mustard, which is to harvest deer, which also significantly reduces the incidence of Lyme. But there is another component of this cycle to consider. Normally, baby ticks (called nymphs) like to feed on mice, which is where ticks are most likely to ingest the Lyme disease organism. After feeding, the baby ticks fall off the mice and overwinter under leaves in the forest. The following spring, they seek out deer (their preferred meal) to feed on before they lay eggs and die. But we do not live in normal times, and overdevelopment has led to crowding of more deer into diminishing acreage of wild fields and forests, at the same time that mice populations are increasing because there are fewer predators to keep them under control. Creatures that eat mice include everything from foxes, hawks and owls to snakes, skunks, and, yes, cats. Dogs and coyotes too. So, if there are too many mice and too many deer and not enough predators, the tick population explodes and then ticks feed off humans as well. What is the solution? Keep the deer population under control with managed hunting. Cultivate and protect mouse predator populations. Which means making sure there are plenty of wild spaces. And then check yourself for ticks, every day.

Is Lyme disease a man-made biological weapon? I think not. The best evidence debunking that story are the plentiful scientific reports that Lyme disease has existed throughout history in humans and animals, even though it wasn't called by that name.

Relax. Worrying about conspiracy theories is not going to make your life better. Or keep you safe. Protect and respect the wild places so that nature can find balance. Let the predators do their work.

Groom yourself. Pay attention to your body.

Groom your loved ones (including your pets). And pay attention to their bodies.

At the end of the day, we can all be tiny nuisances.

But remember, when Tick said, "We are amazing," it wasn't just talking about ticks. It was talking about all of us. *We are all amazing!*

Thank you, Tick.

Aspen

Always use your power for good.

When I first moved from downtown Emmaus to my house in the woods, I spent a lot of time just watching and observing nature all around me. One day, while looking at the trees, I got confused. Were those aspen trees? Don't aspens only grow out West in places like Colorado? I looked it up. Yes, aspens do grow in Pennsylvania.

Next to my campfire pit is a tiny grove of aspens nestled among the white pines. I planted the pines, but the aspens moved in on their own. Their twittering, shimmering leaves dancing in the slightest wind are a pleasure to watch. I had no quarrel with them. It still seemed weird to have aspens in Pennsylvania, but it's not for me to say who belongs here and who doesn't.

Until, that is, the aspens started moving into my garden beds. In just one year, from fall to spring, a dozen aspen saplings had sprung up among my precious Japanese maples, which are one of my greatest garden joys. Therefore, the aspens invading my Japanese maples had to go.

Did you know that aspens clone themselves, and the clones connect to each other underground, creating the largest single organism, by area, on Earth? Good for you, aspens. But you still need to respect my Japanese maples. Please.

I knew better than to remove them without first trying to understand what they might want to tell me. I looked them up in Ted

Andrews's *Nature-Speak* book. What jumped out is that aspens make great "wands" to use to talk to trees. Ha! That gave me the excuse I needed to dig them out. I will make wands.

What does one do with wands? I have no idea. But I can figure that out later.

It was rather surprising how easily they came out, as if they were saying, "Pick me for a wand! No, pick meee!" I dug them all up. But one was a little more difficult than the rest to dislodge. It was nestled in among the sweet William, who I must say really is rather sweet. Suddenly, I saw it — an old, "dead" aspen stump in the sweet William. Ahhh. This stump, which had probably been cut down years ago, had been sending out shoots, children, progeny.

I removed all the little wandering aspens from among my Japanese maples and decided to journey to Aspen to see what else I need to know and understand from it. The bucket of sticks (potential wands) sat by my journeying spot for months. Finally, one day, I felt it was time to see what Aspen had to say. I drank my morning coffee and looked out at the winter view. There was snow on the ground and a pink sunrise. The trees were bare, but I could feel them calling to me. I finished my coffee and got set up. I picked out two aspen sticks to hold onto for the trip.

I opened sacred space, lay down, and started the drumming app.

✱ ✱ ✱

I entered into my tree, and it became a large, dark brown cave — a cathedral of leafless trees. On the far side I saw an opening that led to a beautiful forest of aspens. The light was otherworldly — a dusk blue sky, illuminating vivid green leaves and soft glowing meadowy grass. I walked down a path until I found a grassy clearing and lay down. I felt myself merging with the earth, the trees, the roots, until I became an aspen. My head started shaking back and forth as if it were a fluttering aspen leaf, and I could hear the chatter among all the other leaves. It sounded like giggling and friendly gossip. The glowing white root energy of the trees connected me to everything and everyone — other trees, but also all the people. I could see that, like leaves falling in autumn, when people left Earth (died) they floated out into the universe still connected by the glowing white root filaments. Everything, all the stars and planets and people, were linked.

Aspen spoke:

> *We are all related. We are all one.*
> *We are the pieces, and we are the whole.*
> *We are everything and we are nothing.*
> *The galaxy is a cell.*
> *The galaxy is a forest.*

I could see our galaxy as a cell, with the sun as the nucleus and the outer limits as a membrane. We are all related. We are all relations. We are all one. I could see it. It made total sense. We are inside a cell, and trillions of cells are inside us. So beautiful. A microbiome within a microbiome. A macrobiome within a macrobiome.

"What about the wands?" I asked.

"The wands represent free will. You have power. Everyone has power. But what you do to others, you do to yourself. Always use your power for good. Know that *everything* and *everyone* has power."

Now I was sitting up in the clearing in the beautiful aspen forest. An Indigenous woman wearing white skins, feathers, and furs walked toward me. I couldn't make out her face. She started dancing in circles around me. Three circles. I tried to ask her what she was doing and what she wanted me to know, but she didn't speak. After the third circle she came toward me and knelt in front of me. We held hands, and I looked into her eyes. In her eyes I saw the forests and the galaxies.

"Always use your power for good," she told me. "We are all related. We are all one."

At that moment the drum called me back.

✶ ✶ ✶

I don't know about you, but at many times in my life I have felt pow-erless. Powerless to change things, fix things, solve the problems in the world. During this journey to speak with Aspen, I got a glimpse of a different idea about power: the possibility that every single thing we do, as well as our every thought and every word, holds infinite power and consequences. It's not just the big things—like having kids or pursuing a career—that matter, but the little day-to-day actions, too. Including the words we speak to loved ones. The words we don't speak to loved ones. Our tweets. Even our secret thoughts.

I remember being a little kid and thinking about the idea that "God sees everything" and feeling shame and fear. Everything? Really? Even when I am picking my nose or going to the bathroom? I'm talking about a different kind of visibility. It's not that some old man is sitting on a throne being all judging or shaming. It's the idea that independent of what anyone else thinks or says, every single thing we do has an impact. Everything we do matters.

Lately, I've been thinking a lot about the word *integrity*—mostly because it seems to be lacking in so many people in so many places, all around the world. But I can't control them. All I can control is myself and my own integrity. Do I use my power for good? All my powers? I try. I don't always succeed, but I sure try. Integrity is about being honest, being truthful, and doing the right thing according to an ethical and moral code (of love, I might suggest). Integrity is not allowing yourself to be corrupted or to attempt to corrupt others. We need more integrity in the world. But all I can control is my own. That is my power.

Sometimes it can be overwhelming. There are so many aspects, from thinking about where the foods I purchase come from and how they were created (and where they go when I'm done with them) to thinking about those snarky things I want to post on social media or offhandedly say to family members or friends. I try to be mindful that each of those things has consequences, and the negative conse-quences hurt me too. That *usually* is enough to make me stop and think. (But not always. None of us is perfect.) I try to remember that each action creates a ripple that reverberates throughout the whole of humanity — and the entire universe.

It's up to us to decide how we will use our power, the power that each one of us and everything already has. If we don't use it for good, we are only hurting ourselves. As I've said before, even shamans are not all good. Many people are drawn to spiritual groups, religion, politics, celebrities, and successful entrepreneurs (or billionaires) because they are looking for power for their own selfish purposes, whether it's for fame, wealth, enlightenment, or influence. Many people are drawn to follow those who amass power because it makes them feel special by association and gives them a sense of belonging. But that kind of abuse of power can only harm people in the end. That's why any kind of spiritual pursuit — especially shamanism — must be approached with respect, humility, and caution.

So much abuse of power lies in the pursuit of money. And so much anxiety in the world lies in thinking about who has money and who doesn't, and how to get it if you don't have it and how to get more if you do. While I have been blessed to come from a family that had enough money, and I've also made enough money, I know that money truly does not buy happiness (and it often significantly complicates things). Yes, it's important for all to have enough money for necessities — food, shelter, and a sense of security. But we have to shift our focus from seeking more and more for ourselves to practicing generosity to others. Being generous, as an individual *and* as a government to its people, is the key to ensuring that everyone is safe and secure. There is enough for everyone. And taking care of each other makes the world safer for all of us. Making sure everyone has enough and is safe is not socialism. It's love.

I choose to live life as if we are all related and all one. And I choose to use whatever power I have for good. It makes everything easier. And better. And more fun. I still mess up a lot, but that's just being human, as we all are.

Only love wins in the end. But apparently that's a lesson we need to keep learning over and over. It would be good to remember that while we all appear as individuals, we are all connected with filaments of light, like the roots of the aspen.

We must always use our power for good. Because what we do to others, we do to ourselves. And after all, we are all related. We are all one.

Thank you, Aspen.

Cicada

We need the darkness to grow.

JOURNEY: JANUARY 20, 2022

Every seventeen years, the cicadas known as Brood X, the Great Eastern Brood, come out of their underground homes to fly drunkenly through the air, mate, lay eggs, and die. Their shrieking mating calls sound like alien spaceships. (Or maybe the soundtracks of science fiction movies featuring alien spaceships were based on the cicada call? That seems more likely since we have never actually *heard* an alien spaceship.) For approximately four weeks in summer there is a cicada frenzy of seventeen-year-olds partying like there is no tomorrow. Because for them, there isn't.

While Brood X is native to the northeastern United States, there are cicadas all over the world. As a young woman, I dreamed of traveling to France, and when I finally got there, one of my most treasured purchases was a vintage cicada wall vase I found at a flea market. It is a symbol of good luck and happiness and was one of the first things I hung in my new home, which was built during a Brood X year.

I will never forget standing in the orange sandy soil with the builder, dirt-stained blueprints propped on a makeshift table as we watched the framing go up . . . and being divebombed and bonked in the head by clumsy cicadas on their way to somewhere looking for love. There were too many to count, their reddish-orange eyes more empty looking than scary. (Fun fact: After they die, their eyes turn black.) Cicadas are nothing to be scared of. They arrive, they are a feast for birds and animals and some brave humans (apparently they taste like shrimp), and then they disappear. They don't hurt anyone or anything. But they do serve as a reminder of time passing.

As the Brood X of 2021 emerged, I thought back to what my life was like seventeen years ago and what has happened since then . . . A lot. The house was built, and my family and I moved in. I learned the phrase "new house, new baby," and at age forty-four I had my third child. I wrote *Organic Manifesto*. I became CEO. My mother died. Maya got married. I got divorced. I wrote a cookbook called *Scratch* (testing all the recipes in my glorious new kitchen). I sent Eve off to college. I went to Australia, twice. I became a children's book author (more on this later). I became a grandmother. I sold the family business. I had a studio added to the house for writing and creating. I even survived a pandemic (so far). Whew. All these big things happened while Brood X was doing whatever cicadas do underground, in the darkness, living on their own time frame.

In 2021, a group of fifteen volunteers agreed to spend forty days and nights in an underground cave in France without any technology or timekeeping devices. The Deep Time project, as it was called by its sponsor, the French Human Adaptation Institute, was undertaken to better understand people's natural rhythms. When it was over, many of the volunteers had wildly misjudged the passage of time. Some felt that only twenty-three days had passed. Others thought it was thirty-one days. The mechanism that enables cicadas to live underground for seventeen years and emerge at the same time together is some sort of undiscovered magic, still waiting for us to figure it out. Does it have to do with the magnetic poles? With the moon? With some genetic code? With some internal timepiece embedded in their bodies? Perhaps they aren't individual insects but rather smaller organisms in one bigger "brood" organism?

Here is what I do know from observation: In the areas where we disturbed the land to build our house seventeen years ago, I found no holes from which cicadas emerged in 2021. The relentless hum of their song emanated from the woods around me, but not from right in my garden or yard. I did see cicadas flying into my garden looking for new places to find true love. Not as many as there were seventeen years ago, but enough to know that there is hope . . . except for the poor cicada my cat ate (and then promptly threw up).

A new generation of cicadas will emerge in 2038. If I'm still alive, I will be seventy-six. Who knows what will happen between now and then? We humans are an odd species, often imagining the worst. Dystopian science fiction films imagine futures of darkness and devastation, overcome by robots that look like insects and larger-than-life evil villains. The film *Blade Runner*, which was made in 1982, the year my first daughter was born, takes place in Los Angeles in 2019 and imagines it as dark, constantly raining, and apparently overrun by flying cars. (Time is not kind to some movies.) The *Matrix*, made in 1999, with its famous red pill/blue pill scenario, imagines a future where machines that look like insects harvest humans for energy. Both realities are dismal, and both movies are filled with slow-motion gun fights, hand-to-hand fighting, and life-or-death moments with ridiculously and intentionally scary scenarios. Personally, I don't understand why people enjoy watching these sorts of things, but I am probably in the minority.

Perhaps in 2038 we will all be dead from nuclear annihilation, climate change, diseases, or some toxic environmental catastrophe. Or maybe humans will still be here, living life from day to day, doing the stuff we humans always do — eat, have sex, have kids, go to school, work, party, enjoy our families, shop, watch shows on whatever technology will be invented to entertain and distract us. Live. Die.

Most likely, life in 2038 will be a mix of good and bad. After all, the real-world Los Angeles in 2019 was still sunny and semifunctioning, even if it was overrun by cars — but none of them flying. I like to think that we are a very creative species, always coming up with new solutions just in the nick of time. Perhaps in 2038 we will finally incorporate our new understanding of how the carbon cycle works into how we grow our food and manage our environments. Perhaps we will have transitioned away from fossil fuels and rebuilt communities around quality of life rather than cars. Hopefully we will have found a new way to travel from place to place that doesn't involve car tires. Perhaps we will have found the courage to build a new financial system that mimics the cycles of nature — birth, growth, death, and rebirth — rather than the exhausting and unsustainable expectation of constant perpetual growth. Perhaps we will learn to respect and honor diversity of all kinds, rather than vilify and fear it.

Perhaps we will consider the cicada.

It doesn't bother anyone. It doesn't bite. It does only minor damage to the trees where it lays its eggs. It's called "flagging" — the end of a branch droops over and dies after the cicada inserts its eggs into the branch. It's not a big deal for the trees. It's more like a gentle pruning. When you look closely at a cicada, you can see beauty, good luck, happiness. Cicadas are both fleeting and constant — visible to us for only a few weeks every seventeen years, and otherwise safe underground, minding their own business.

But of course, me being the nosy, curious woman I am, I decided to pay them a visit to see what's *really* going on down there . . .

On the day of my journey, it was a beautiful snowy morning. I decided to use my Australian sticks to wake up the spirits instead of my drum. I noticed that I was starting to feel so much less inhibited about opening sacred space. Ever since my weather journey, I'd been using my body more freely — yes, dancing. It felt so good to move my body in whatever way I chose to call in protection and guidance. I was moving for me. No one else.

After I opened sacred space, I lay down and began the drumming app, asking Cicada to help me understand it.

* * *

I walked into my tree. A noise came out of me, sounding like a cicada call. It lasted for about a minute — one breath. I jumped into the abyss. It was dark. Really dark. Duh, I thought to myself, I'm underground where the cicadas are. Of course it's dark. I felt activity around me, baby cicadas going about their business, carefree. Then I heard a voice (in my head): "Time is an illusion. We need the darkness to grow." I knew right away that Cicada wasn't just talking about lack-of-light darkness, but about the trials and tribulations we humans experience in our lives.

"The darkness underground is our Middle World. Your Middle World is our Upper World." I sensed a universal law of scale — that we experience life as scales within scales based on our size and perspective.

I felt myself "rising up to the light," and the experience was like descriptions I've read of near-death experiences. Except I started to get very, very hot. So hot I had to throw off my blanket. I was transitioning from the coolness of the soil to the heat of summer sunshine. Then I was out in nature flying about with joy. It was like being born.

Your world is our heaven — our reward. We celebrate. We find love. We mate. We die so we can be reborn through our children. But we all need the darkness to grow. There are many kinds of darkness:
> *Lack of light*
> *Lack of knowledge*
> *Lack of wisdom*
> *Lack of love*
> *Your job is to grow from darkness to light, from ignorant to knowing and wise, from hate and fear to love.*
> *It's all part of the dance of life!*

I made the mating call of a cicada (though it's only the males that make the call). I felt myself flying and mating. I watched my mate fall to his death. I felt myself fill with eggs. I laid them in a branch, my lower body alternatingly scissoring open the branch and thrusting in the eggs. As I fell to my own death, I was swooped up by a bird. It might have been a vulture, because suddenly I was with a vulture, and I asked if it wanted me to journey to it — this time without fear. It nodded its head yes.

I went back to Cicada to thank it and ask if it had any other messages for me. It repeated: "Time is an illusion. We all need the darkness to grow."

✳ ✳ ✳

Is time an illusion? Yes, according to many physicists. Einstein said time is relative. Others have come up with all sorts of theories, from the existence of multiple dimensions happening simultaneously to time being completely fabricated inside our brains. But no one really knows for sure.

There's a lot we don't know. For example, scientists believe the universe is primarily made up of an unknown, unreactive substance called dark matter. What is it? No one knows. It's a mystery! (My guess is that it is love.)

At this point I should probably tell you that I'm a believer in the value of science and modern medicine. I got the COVID-19 vaccine and booster shots. But as an avid reader of history and current scientific and medical news, I know that science and medicine are fallible. Both are a continual process of experimentation and trial and error, rife with the human emotions of doubt, ridicule, skepticism, subterfuge, and competition. We would not have gotten this far (assuming that time is at least partially real) without the pioneers, explorers, investigators, and journeyers to chart new paths and discover new things.

I served for a few years on the board of directors of a major non-profit hospital, which gave me an understanding of how the medical system works. And my career in health publishing taught me how advertising, and especially pharmaceutical advertising, works. Because of these experiences, I understand why many people's trust in science and medicine has been broken. I've also seen how funding has skewed scientific research for the benefit of the funder, whether it's a chemical, tobacco, or sugar company (or political party). These kinds of shenanigans are happening all around the world, no matter what the political or economic system. Our society's current lack of trust in all our systems — scientific, medical, and political — stems from the damage caused by human greed, ego, and desire for power and domination that can never be satisfied. It's the Sackler family finding new ways to sell opioids to boost their profits so they can put their names on museums. It's Rupert Murdoch and his son attempting to bring down democracy by creating an empire of fake news–fueled outrage so they don't have to pay taxes. It's Vladimir Putin invading Ukraine to satisfy his own personal agenda of (false) Russian supremacy and to defend an imaginary idea of "family values" that he himself doesn't practice. It is humans. Crazy messed-up humans.

We are the source of all the misery. But we can also be the source of true healing. We need the darkness to grow, right? To force us to find the light — the light of knowledge, wisdom, and love. I am reminded of the yin-yang symbol: black and white, negative and positive, left and right. Which brings to mind a college essay in which I equated the left and right in politics to the human brain. We need both left *and* right brains to make a whole brain. We need the darkness *and* the light to survive and grow. We need the negative to propel us toward the positive.

The darkness is real. We need it to grow. But the light is also real. All any of us can do is stretch ourselves to learn, grow, build trust with those around us, and live each day in the present moment. And trust the magic . . .

During final edits of this manuscript I was looking up an article that I remembered reading online in *Emergence Magazine* to fact-check something in another chapter. The very first article to pop up was titled "Chasing Cicadas," and, weirdly enough, it had been written by Maya's best friend from high school, Anisa George. As I read this incredible piece, its beauty and bravery made me cry. But that wasn't the weirdest part. I had never looked up the Latin name for the Brood X cicada, and from the article, I learned that the genus is *Magicicada*. MAGIC-icada! Anisa informed me that the name is actually pronounced *magi-cicada*, and it derives from *magi*, meaning king. But I still think it's funny, weird, and totally magical. I just love this crazy universe.

Whatever happens between now and 2038, it's almost certain that the *Magicicada* will emerge again. And I will feel quite fortunate if I am there to greet them.

Thank you, Cicada.

Groundhog

It's OK to be normal.
Everything is normal.

O h groundhogs, you are such an interesting part of a gardener's life — and especially a Pennsylvania gardener's life. After all, we are home to Punxsutawney Phil, the groundhog who sees or doesn't see his shadow on February 2 to determine how much longer winter will last. The Inner Circle — a group of local dignitaries — gets ready for the day by preparing two different scrolls with different outcomes. In the early morning of February 2, they wake the groundhog from his bed on Gobbler's Knob, and the rodent speaks to the president of the Inner Circle in "Groundhogese" to let him know whether or not he's seen his shadow. The Inner Circle are all men dressed in black overcoats and top hats. But only the president is able to understand Phil, who is still a bit delirious from spending a few months in the dark sound asleep without eating or drinking anything. The scroll with Phil's answer is read aloud to thousands of "phaithfphil phollowers" — many who have been up all night drinking and waiting in the cold to hear the news, which is then broadcast around the world. Did he, or didn't he? That's what everyone wants to know.

See, I am not the only weirdo who talks to animals. At least in Pennsylvania.

Groundhogs can be a destructive nuisance because they are voracious and love to eat vegetables. They also are epic diggers. The more I dig into groundhog history, the farther back in time I end up, discovering that they tell an important story of how we have traveled *through* time (even if it is an illusion), picking up bits and pieces of culture all along the way.

Let's start with February 2, Groundhog Day. Why February 2? Well, that is the day the Christians celebrate Candlemas, a holiday that commemorates the presentation of Jesus to the Temple, and the purification of Mary. Basically, according to the Bible (Leviticus 12) a woman was to be purified in the temple thirty-three days after she gives birth by offering a burnt lamb sacrifice and a young pigeon or dove as a sin offering. This was to cleanse her from her postpartum bleeding. When Jesus was born, Mary brought him to the temple (Luke 2:22) and offered a pair of turtledoves or two young pigeons (not presenting a lamb meant that they were unable to afford it, or it was no longer required — times change).

But why do we celebrate Candlemas on February 2 and what does the groundhog have to do with this? Candlemas also happens to fall on the date of the Pagan celebration called Imbolc, which means "in the belly" and refers to when animals get pregnant in the Spring. Imbolc originated as a Celtic festival to mark the midpoint of the season between winter solstice and spring equinox. It was also celebrated as St. Brigid's Day. Before Brigid was a saint, she was a pagan goddess renowned for her generosity. Her special day heralded the coming of Spring and involved weather divination, as described in an old British rhyme:

> *If Candlemas day be fair and bright,*
> *Winter will have another flight.*
> *If Candlemas day be shower and rain,*
> *Winter is gone and will not come again.*

The Germans, who were also originally pagans (as all people in Europe were before the advent of the Abrahamic religions), first used a bear to help predict the weather because this is around the time bears come out of hibernation. When bears became too hard to find, badgers, and then hedgehogs, were used instead. Then came the Catholic Church, which turned these pagan celebrations into Christian ones.

But still, I keep digging. How did this pagan groundhog ritual get to America?

You can blame it on the printing press. Thanks to the printing press* being invented in the 1400s, which enabled people to read and interpret the Bible for themselves, the Catholic Church, afraid of losing its complete authority to translate the Bible to the masses, fought back by murdering the protesters, aka "protestants." *That* is why so many Protestants immigrated to North America — to find religious freedom from the Catholic Church, the Church of England, and all the fights among them.† Along with their religious beliefs, the new immigrants to America brought their pagan traditions. The German Protestant Lutheran immigrants settled here in Pennsylvania and are now known as the Pennsylvania Dutch, which is a version of the German word *Deitsch*.

This massive migration for religious freedom also included the Pilgrims (who were Puritans), Moravians, the Amish, Quakers, Mennonites, and also the Jews (the other side of my family). When the Pilgrims and others got on a boat and sailed to an unknown land with dandelion seeds in their pockets, they were looking to find a safe place to worship God as they pleased without getting murdered *by other Christians*. Sadly, those same European colonizers did a lot of bad things in their escape to find freedom. They attempted to exterminate the native populations. They brought in slaves to do the most

---

* With every major introduction of new technology, the existing social order gets disrupted — from the printing press to the automobile to TV to the internet.

† It was violent Christian-on-Christian violence that the Founding Fathers wanted to guard against by separating the church from the state.

difficult and dangerous work. They stole land. Each successive wave of immigrants to North America fought among themselves and against every new group that came after looking for the same opportunity, the same freedom. But in the slow process of integration, this diverse blend of people brought many traditions from their original countries into their new lives, where we often take them for granted today.

That's where groundhog comes in.

Since there are no hedgehogs in Pennsylvania — or anywhere else in North America (except perhaps in zoos or as pets), groundhogs boldly stepped into the breach to help celebrate St. Brigid's, Imbolc, and Candlemas day.

Groundhog Lodges (*Grundsow Lodsch*) became men's groups in America aimed at preserving Pennsylvania Dutch language, culture, and odd sense of humor. But both men and women were the "shamans" of PA Dutch culture. They culturally appropriated the word *Powwow* from the Algonquin language, because they believed it referred to the same type of healing work involving trances, dreaming, and divination that they had performed in the Old Country. And while their healing work was deeply rooted in their Lutheran Christian faith, the origins go back even farther to the folk healers that later became demonized by the church as "witches."

The pagan/heathen pre-Christian roots of Germanic culture can still be seen today: Hex signs on barns and mirrored balls in gardens. I've seen mirrored garden balls all over Pennsylvania for my whole life, but now I understand them. They apparently ward off negative energy and attract good luck.

I planned to wait until February 2 for my journey to talk with groundhog. But on January 27th I got the urge and was ready. I asked for a sign from the groundhog and started my morning meditation. Right away, a big groundhog appeared to me. But this guy was funny! He was flopping around like he was exasperated with waiting for me. He drummed his fingers on the ground. He rolled his eyes and puffed his breath and crossed his arms. OK, OK! But I still had to finish my meditation (twenty minutes with a timer) and open sacred space.

* * *

Once I lay down and started the drum, Groundhog was right there in his den. Weirdly, there were lights blinking behind my eyes along with the beat of the drum. (It felt a bit like an old movie.)

"Sheesh what took you so long? I'm getting antsy and rutchy. (*Rutchy* is a Pennsylvania Dutch word that means squirmy, jumpy, and just plain annoying.) You woke me up early. I'm starving. I can't wait to get out there and eat fresh green things." (Sadly for him, there was still snow on the ground outside.)

I handed him an energetic bunch of carrots, which he ate with delight. Then he started to fall back to sleep.

"Wait!" I cried, "What is your purpose?"

"We don't need no purpose. Why does everything need to have a purpose? We eat. We sleep. We hang out with family. That's all that matters anyway. Not everyone needs to save the world. It's OK to be normal. Most people just want to be normal."

His language and way of talking was confusing to me, because he sounded belligerent and dismissive. And maybe a little bit Pennsylvania Dutch. Most of the other beings who had spoken to me during journeys had been articulate and insightful. I suspected that Groundhog's manner was some sort of act. But I wasn't sure why and what would come next.

What came next was a bit like an animated Robin Williams take on the Groundhog mimicking humans. He put on some wire-rimmed glasses and stuck a pipe in his mouth.

"Some people think they are so important but they're not."

Then he put on the top hat and black overcoat like the Punxsutawney Phil guys.

"Some people look ridiculous, and they are."

Then he put on a sexy red bra and got into a suggestive pose, his brown furry belly hanging out over tiny red panties.

"Some people think sex will make them happy, but it doesn't."

What is normal? I asked him.

"Everything is normal." Now he was speaking like he was from New Jersey. "Look, we got the gays down here. We got the transgenders. It's all normal. Stop worrying so much about everything. You guys worry way too much. Just live within your means. Here's what you should worry about: eating, sleeping and who you let into your den. Oh, and the occasional good time." He put on a Groucho Marx mask and had a cigar in his mouth. "If you know what I mean . . . Hubba-hubba!"

At this point I was not exactly laughing aloud, but I was definitely impressed and entertained and in awe of his acting ability. Then he struck a very serious meditation pose, but he opened one eye to see if I was watching him. He was making fun of me!

"You all need to stop taking yourselves so seriously. It's boring." He said dryly.

And then the drum called me back. I tried to give him a goodbye hug, but he recoiled in horror. "Don't touch me!" He shrieked as if I had cooties. I laughed, thanked him, and left.

* * *

An hour later I was still laughing.

I've never really had a problem with groundhogs in my garden, by the way. They live in my front yard in a huge rocky Groundhog Condo Complex. It is truly hysterical in the spring when the babies, called chucklings, come out and if you drive by them, they freeze as if to make themselves invisible. If you stop the car and open the window they will sit there still as statues until you get bored and drive on.

But I do know others who struggle with groundhogs in their gardens, and I suspect the solution is the same as rabbits: high raised beds and metal fences. Plenty of predators. And plenty of wild spaces to get their fill. They really prefer to eat "weeds" like dandelion and clover, which is why I often see them happily chomping in my front lawn.

Some people hunt and eat them, but not that often anymore. (Although originally Punxatawny had a September groundhog hunt as well, which culminated in eating a lot of groundhog pie.)

One of the greatest letter-writing battles in magazine history arose after someone wrote to *Organic Gardening Magazine* recommending that gardeners shoot, kill, and eat groundhogs to prevent them from stealing food from the gardener (I think there might have even been a recipe for stew). It was rather emotional between the animal lovers and vegetarians versus the practical, meat-eating hunters and serious gardeners. Fortunately, no one got hurt. Except maybe a groundhog that was hit by a car while trying to cross the road. Sheesh people, lighten up. If you are going to garden, protect your space. And, by the way, also respect the space of others to practice their own faiths. Even their own food faiths.

Everything is normal.

And it's OK to be normal.

But also, it's OK to not *feel* normal. That's normal!

Thank you, Groundhog.

Milkweed

*We are all connected by
the thread of love.*

JOURNEY: JANUARY 28, 2022

Imagine an absolutely perfect Pennsylvania June day. My favorite time of the year. The sun is shining so brightly that all the plants and trees are glowing. Solstice is a few days away, and sap is still rising in the plants. The bees and dragonflies are reveling in the nectar of nature. I can feel it — the humming and buzzing of life at its peak. I am swimming gentle laps in my pool, the turquoise sparkle of water one of my greatest pleasures. I reach the deep end and rest against the edge, enjoying the beauty of summer's glory. There, blooming like fireworks before me, is milkweed. I can smell its heady fragrance. I didn't plant it, it just grew there beside the pool. If I were nature's designer I couldn't design a more gorgeous summer flower — the most subtle rich pink blossoms in big balls that bounce in the breeze, and elegant thick green leaves that stand straight and compliment the flowers perfectly. Can you believe I once considered it a noxious weed?

Yes, I owe Milkweed an apology.

I wrote my first book — a gardening book — more than twenty years ago (1997). It was an exercise in keeping my sanity while going through some personally challenging times. But I also wrote it to show that organic gardens can be beautiful, because back then *organic* had a reputation for being... let's just say, more focused on *functional*. I have always loved to create landscapes and gardens, but there is so much to learn that even the oldest and best gardeners in the universe don't know the half of it. Back then, I didn't have a personal relationship with Milkweed, and no one seemed to be overly worried about monarch butterflies. They weren't on my radar. My gardening knowledge came from reading old books, magazines, and newspapers and talking to farmers and other gardeners and landscapers. And back then, all the farmers agreed that Milkweed was a noxious weed. The *worst*. That's why I wrote that in my gardening book, where it remains to this day. (Fortunately, out of print. But it was a labor of love and still available on Amazon, by the way.)

I'll never forget the first time I saw milkweed in real life. Have I mentioned that I am a plant pig, always on the lookout for new plants to stuff in my garden? I was at the Rodale Institute, and there was one small garden bed filled with tall, elegant plants with loads of large round pink flower blossoms. It was simply gorgeous.

"What's that?!" I asked the gardener.

"Milkweed," she said flatly.

"I thought Milkweed was a noxious weed?" I said.

"It is for farmers," she replied. In fact, at that time a farmer could be fined up to $100,000 if he let it grow in his fields.

In those days you couldn't buy milkweed at a nursery, so I forgot about it.

Fast forward to the 2010s. Everyone is talking about the decline of the monarchs. I hadn't realized that milkweed was the *only* plant that monarch caterpillars ate, and the only plant that the butterflies lay their eggs on. I was working in my brand-new garden, and I don't remember how it happened, but there it was. I know I didn't plant it. I'm not sure why I hadn't weeded it out before then. In retrospect, perhaps it's because milkweed is one of those plants that doesn't look weedy when it first comes up. I usually let plants like those grow to see what they become.

And it became . . . milkweed!

I was very glad because I had never seen a monarch butterfly in my garden, which deeply concerned me. I let the milkweed grow as much as it wanted to. It's common milkweed, *Asclepias syriaca*.

Still no monarchs.

The next year I let it grow again. It spread.

Still no monarchs.

The next year I let it grow again. It spread some more.

Still no monarchs.

I was starting to lose hope. But I had learned to love the milkweed. I loved the scent the most, and the beautiful flower. While it continued to spread throughout my landscape, I appreciated the effortless beauty it brought to my garden. And at the end of summer, I enjoyed watching the feathery milkweed seeds float up and sail away to find even more places to grow.

Then the fifth summer it happened. I saw a monarch!

A caterpillar even built its chrysalis on the back of my campfire chair right where I wouldn't miss seeing it. It felt like a thank-you gift of the very best kind.

Every summer now I see at least five. I like to think word-of-proboscis (that's the name for a monarch's mouth) has spread the news that my place is a good place. Plenty of milkweed to raise babies, and flowers to drink nectar afterward. And no nasty chemicals, either.

In October 2020, I was telling a visiting friend about the monarchs and milkweed and how the last generation of monarchs each summer (there are four generations) flies all the way to Mexico for the winter. Just then, we looked up and saw a monarch flying a little erratically above us as if to say "Wait for me! I'm coming!"

"You can do it!" I yelled up in encouragement.

You can do it!

I hope it did.

One of my very first shamanic journeys involved the monarch. When I was at Esalen in 2014, someone mentioned that monarchs were overwintering in a tree on the grounds (West Coast populations of monarchs overwinter in California, not Mexico). I looked and looked and finally found the tree. It's a crazy magical sight to see, hard to spot, because when monarchs are hanging in a tree with their wings folded up, they look like brownish leaves. But once you figure that out, you really *see* it. Mind blowing.

Here's what I wrote in my journal about my shamanic journey after seeing the monarch tree:

> *I turned into a fairy and flew up to meet with them as they all congregated together on the tree. There were thousands and thousands of them! One of them flew me over the landscape and all I saw was shopping malls, concrete, barren suburbs, and weedless farms. They told me that since the wildness is gone, there is nothing for them to eat and lay their eggs on and they are thirsty. They also told me that all of them together — generation after generation — are all "one mind."*

What does it mean to be all one mind? At that time, I wasn't sure. Some spiritual traditions have explored the idea that consciousness exists outside of us, and our brains are both observers and participants, and possibly creators too. Doctors such as Larry Dossey and physicists such as Bernardo Kastrup have also explored these concepts. They postulate that consciousness itself is everything all together. If consciousness is the ocean, we each are waves. If consciousness is the universe, we each are stars. Honestly, I'm still not sure what to believe.

But I was sure, after that workshop and learning how to journey, that I had found the path I wanted to travel on. And here I am, writing this book, which is a culmination, a flowering, and a metamorphosis from whence I began. And you have traveled along with me here. Thank you. There is one more journey to take — to Milkweed.

✳ ✳ ✳

I dove into the blackness of the hole in my tree with the confidence of Hiccup from *How to Train Your Dragon* (the later movies). I landed in a gorgeous meadow of Milkweed — the tall, mauve flower heads swaying in the gentle breeze, the fragrance filling my soul with joy.

Mother Milkweed. Of course. I immediately knew her as Mother Milkweed.

"We are love made visible," she said.

Suddenly I was swirling around, lifting off the ground, and turning into a Monarch. I drank from Milkweed's sweet nectar. I laid an egg on the underside of one of her leaves. I became a Monarch caterpillar and ate her delicious crunchy leaves. Then I left to become a chrysalis and turned into a butterfly again. Yes, she is the mother!

I became one of her seeds, which are fluffs that float through the air with a flat seed as the stem. I sailed up, up, up into space and then into the rainbow upper world where a dozen laughing spirit children were jumping, playing, and trying to catch me. One held on and together we sailed back to Earth where my seed was planted and grew and gave shelter to the little spirit child while it waited to find a mother. When a human woman appeared (a slender brown-haired woman in a slim skirt and a button-down sweater) the spirit baby jumped onto her shoulder. I saw a fetus growing inside the woman until it became like a chrysalis, waiting to be transformed. When the mother gave birth, the spirit baby entered the real baby on the child's first breath. I felt it happen . . . like a little . . . hiccup!

Milkweed spoke again: "We are all mothers." I saw mothers holding hands together around the world, like the roots of the Milkweed. There were a few men there as well. She heard my questioning. "Men can be mothers too. Not physically, but in their hearts. One doesn't need to give birth to cultivate and create the future. We must all become the mothers and parents to all the children. Creation is love made visible too."

Then I saw her — a Goddess with a crown like a golden milkweed flower branching out all around her head like a fireworks halo. She was surrounding the whole planet with her love. I felt myself being pulled up into space to settle into a warm glowing sphere. As I watched the goddess from afar, I saw glowing butterflies pour out from her, all connected by white shining threads of light.

"We are all connected by the thread of love," she said.

And then the drum called me back.

* * *

Reader, I wish you could feel what I felt in that moment. I wish I could have stayed longer to just experience those feelings. But I had to dash to the supermarket because it was snowing, and I needed bread (what are the odds?). It felt strange to embark on such a mundane task while I was still reeling a bit from the journey — both because it was so beautiful and because it was not what I had expected. And it was the last journey on my list for this book.

Of *course* she is Mother Milkweed. How did I not see that before? Perhaps that's why I feel so connected to her.

For many years, especially while I was blogging on a regular basis, people would ask me when I was going to write a parenting book. I avowed never to do that since being a mother feels like being on the verge of failing almost every day — no matter how old our kids get. But I was inspired to make a list called The Parent's Creed to articulate what I have found to be the keys to raising happy, loved children. Because I do believe that raising happy, loved children is one of the most important ways we can create a better future. Here is my creed:

THE PARENTS' CREED

We are all mothers. We are all fathers. We are all parents.
All children are our children.

Our job as parents is to nurture and help children grow into
who they truly are, not who we want them to be.

We will raise our children without shame for
their bodies and natural desires.

We will protect and not intentionally harm our children
in any way — physically, emotionally, or spiritually
(however, no one is perfect). And when we feel we can't cope,
we will ask for help. There is no shame in asking for help.

*We will teach our children to be resilient and independent —
physically, emotionally, and spiritually. And the best way
to do that is for us, as parents, to learn to be resilient and
independent. We will let them fall and fail over and over
again, and encourage them when they rise back up.*

*We will teach our children to be responsible,
to care for others, to be kind and generous.*

We will encourage and enable our children to marry for love.

*We will teach our children that sex is a healthy part of
a loving relationship, and we will model that by being and
staying in a healthy relationship ourselves. Or having the
courage to leave an unhealthy relationship.*

*We will not look to our children to fill the emptiness
in our adult relationships and substitute their love
for mature, grown-up love.*

*We will find the courage to acknowledge the hurt we cause
our children, demonstrate the power of apologizing,
and make amends when possible.*

*We will give our children the freedom to explore who they are
and what they want from their lives, without censor or control.*

*We will love our children, and all children, as unconditionally
as we possibly can, and always welcome them home.***

---

* I realize that some parents will not agree with all the points in this creed.
But after raising three daughters in three separate decades, (from MTV
to Hannah Montana to Snapchat) I feel confident to share what has
worked for me.

During the Milkweed journey, seeing the spirit child seeking out a mother and entering into the body of the baby as it inhaled its first breath reminded me of my own story. I bore a child out of wedlock when I was twenty. I *chose* not to abort the baby. I knew my parents would help me and my child if we needed it, and I also knew this child had come to me for a reason. I was so glad I had the choice. And I made it for myself. (There were many people who wanted me to have an abortion, including my mother.) Being a single mother in 1982 was not easy. I saw how the world was geared almost exclusively toward men. Children were seen as a necessary nuisance for women to take care of — preferably out of sight of grown-ups. And yet as a single mother it was up to me to support my child, which made my drive to work and succeed primal. (All the other successful women in business I met at that time were also single mothers.)

Things have changed since then. A LOT. But they have also not changed. Raising children is still primarily a woman's burden. And a hard one at that. Becoming a single mother at such a young age, and in such a pivotal time for women, taught me so much. But the most important thing I learned was that love is the *primary* thing that matters. It didn't matter what people said about me behind my back or gossiped about me at work. It didn't matter what my mother's friends thought and said to each other. It didn't matter what I wore or how much I weighed. It didn't matter what my daughter packed in her lunch to school (sardines and Cheerios were favorites). What mattered was the personal relationship of love and understanding that I created with my daughter, who is now a grown woman with two incredible daughters of her own.

Fathers are super important too. But sometimes they just aren't there. And sometimes even when they are there, they aren't *present*. Both mothers and fathers play an essential role in raising happy, healthy children, which is why we need to learn how to do that for the best outcome for our children. And why I call it a Parents' Creed, rather than a Mother's Creed (which was my original title for it).

There is something about the "pro-life" and "pro-choice" abortion debate that reminds me of my original encounter with Mugwort.

On both sides, people are banging their heads against a brick wall over and over again. How can we shift our perspective? Can we look at sexuality, love, and motherhood with new eyes that are open to considering different paths forward? Life is sacred. *All* life is sacred. *Freedom is sacred too.* I suspect if we put all our energy into finding a new path forward, we can find a way to care for babies, mothers, and all people in a better way — a more loving way — putting all the teachings from nature into practice and bringing us all into balance.

Nature *needs* us to learn how to love. And our children and their children *need* us to learn how to love. We already are love made visible. We already are connected by the glowing threads of love. We just have to open our eyes to see and, more importantly, open our hearts to *feel* it.

What do we need to feel?

Love.

Nature.

Magic.

Once when I was planting something, I accidentally dug up the giant, white, underground runner root of a milkweed. It was so thick and strong I couldn't believe what I was seeing. It was the size of a garden hose, but pure white. Milkweed spreads from a mother plant (of course — I only learned this after my journey) that sends out runners from which other milkweeds sprout up. This underground root is like our power, our thread of love, our connection to each other and to the Earth. It's there whether we see it or not. It's always been there. Now it's up to us to tap into it and heal our hearts, because that's where the healing needs to start. We can hold hands together and create a force stronger than any hatred. Your heart. My heart. Our hearts. That is our work here together.

The thread that connects us cannot be broken. We rise and fall together. Healing the human heart is what will heal the planet.

Nature is waiting.

Thank you, Mother Milkweed.

The Fungus Among Us

Sing the world into being.

JOURNEY: MAY 21, 2022

The manuscript for this book was in final edits, and I called Lisa for some last-minute fact-checking. She asked me how I was doing.

"Fine," I said. "But not really. I must have picked up some ringworm from my garden and I can't seem to get rid of it." (Ringworm is a common fungal infection, not an actual worm). There was a red round circle the size of a nickel on my calf. I'd been to the dermatologist and ruled out anything truly deadly, and I was faithfully applying an antifungal cream twice a day. But the red circle wasn't going away.

"Maria," Lisa reminded me, "you didn't include Fungi in your book, did you?"

In truth, the moment I hit send on the first draft to my editor, I had had a moment of panic because I hadn't included any fungi, and they are so important to ecological systems.

"No! I thought about it but couldn't decide on which fungi to focus on. It seemed so overwhelming." Also, fungi had never really annoyed me — until now.

"Well, now you've got your answer, don't you," Lisa said with a grin so bright I could see it through the phone lines. "Go talk to them! And don't forget to ask the ringworm to leave your body." She felt pretty certain that the ringworm was sent as an emissary to pester me into including the fungi in my book.

I recalled that it had rained the night before my phone call with Lisa, and just outside the window beside my writing desk, I could see that about fifty mushrooms had popped up in the grass. Yes, they were trying to get my attention.

I flashed back to a journey a few years ago when I asked permission of the land to build an addition on my house, creating the space where I now do my writing and journeying. During the journey all the nature beings sat around a conference table discussing how I was to honor them with my work. It was funny and sweet, and I remember a baby mushroom sitting there looking much like the adorable mushrooms in George Lucas's underappreciated movie *Strange Magic*.

Now, it's a Saturday morning, the morning after my call with Lisa. And I'm going to journey to speak with Ringworm and whoever else shows up from the Fungi realm. The weather is still perfect for mushrooms to sprout — warm, foggy, damp. The dew is sparkling in the morning sun.

I walked toward my tree. The earth felt spongy. Before I could climb into the opening the ground absorbed me and I went under. I immediately started singing a Christmas carol:

> *Angels we have heard on high,*
> *Sweetly singing o'er the plains,*
> *And the mountains in reply,*
> *Echoing their joyous strains*
> *Glo-o-o-o-o-o-o-o-o-o-o-o-o-o-o-ria, in excelsis deo . . .*

I cried. Because I always cry when I sing — especially hymns. (The French word for this is *chantepleure*. It's a real thing.)

"We sing the world into being," I heard the fungi say.

Then, I heard more music. Other voices, not me, were singing that song from the original 1966 *How the Grinch Stole Christmas!* television special. The one where the Whos in Whoville hold hands and sing "Falu, Falu, Falu . . ."

Suddenly I just knew . . . We need to sing! Singing is our power to create the world!

Then I felt like I was swimming underground.

"We are the ocean under the surface of the Earth. We decompose, connect, rise briefly to reproduce, and spread our spores and then disappear under the Earth again." I saw the puff of spores spreading from a mushroom above ground, just the way a whale or dolphin expels air and water through their blowhole.

"You are fungi too," they said.

I asked about my ringworm infection.

"Stop thinking so much about death. You have been thinking too much about death. You are not ready to die. You need to live."

"I want to live!" I cried.

"Then sing!" I saw people singing hymns in churches (all kinds of churches). I saw rock concerts. I saw English football fans singing in unison at their games. I saw Indigenous people singing around fires. "Sing the world into being."

Then an *Amanita* mushroom appeared, the iconic magic mushroom — red cap with polka dots and a white stem.

"You don't need to eat us to know our wisdom."

What is your wisdom? I asked.

"We are magic. You are magic. The universe is magic."

I saw a person with the *Amanita* mushroom tattooed on their back. They lifted their arms and their arms turned into white wings and they flew away.

I asked the ringworm to please leave my body, and I felt it receding. Then a being of white light came and swept my body clear.*

The drum called me back.

---

* Readers, the red ringworm circle on my leg immediately started to fade after my journey, and a week later was completely gone. I still applied the antifungal cream just in case.

✳ ✳ ✳

Wow. Whoa.

The truth is, I have been thinking about death too much lately. I just turned sixty, the age when my father was killed. And I have been obsessively attending to my house to be sure it is in full working order and upgraded "in case" something happens to me, or to the world, or simply to be prepared for getting old. I am not afraid of death. But I am always an advance preparer (my kids would say I'm an excessive over preparer).

It feels like a tremendous shift in my life — there is death happening — the death of being a mother with kids at home (which I've been doing for forty years!). Even finishing a book manuscript is a kind of death. I am aware every day that I am now in the last decades of my life. It's a very good life, and I am focusing on what keeps me healthy and strong, and what brings me joy. I am excited to focus even more on what I need, and what I enjoy.

Now I know I need to spend more time singing!

But Christmas songs!? It's almost summer! I looked up that song from the Grinch TV show, and the words of the song are actually "fahoo fores, dahoo dores." What does that mean? I have no idea and there are no obvious translations, but the name of the song is "Welcome Christmas." And the important message of that song and the original Dr. Seuss story is showing the Grinch that Christmas isn't about getting gifts, or eating roast beast. It's about love. And the Grinch's heart grows bigger, just like mine did while writing this book. (And hopefully yours did, too, as you read it.)

I do think the *Amanita* mushroom wanted me to tell you about its role in the creation of Christmas. *Amanita muscaria* is a hallucinogenic mushroom used by the Sami people — the Nordic shamanic people who herd reindeer. Many believe that the origin of the story of Santa and his flying reindeer come from this tribe and what they learned from ingesting this iconic mushroom. (There is a great little book about it called *Santa Sold Shrooms*, by Tero Isokauppila.)

Isn't it funny and wonderful how all mixed together our traditions are? Christmas is both the birth of Jesus *and* Santa. All the festivals of lights that occur around the world in the darkest part of winter speak to the power of light to overcome darkness. I will light a candle in any church or temple that allows me. And I'm a big fan of Jesus. I'm an even bigger fan of all the Mary's — the virgin, the Magdalene, and the Black (and white) Madonna. And I love Santa too. The older I get, the more I have sought to understand my Jewish roots as well. Jews place rocks on the graves of their loved ones in the same way that people throughout history have placed stones in piles on mountains and pilgrimage sites to honor the site. "The rocks represent prayers," Lisa explained to me. "They're located in power spots. Often people stop to offer prayers of respect for the Spirits of Place (and also the overarching Spirits), for safe passage, and often to simply express gratitude." Lisa says in Tuva, they are called oovahs. And yes, that rhymes.

During my journey, I saw that while above ground we all seem different, we are all connected underground. In a way, all living beings are just different varieties of one species — we surface (are born) and live and reproduce, then we die and decompose and disappear underground again. All working together. Celebrating the same things in unique ways. Connected. All part of the same ocean. All part of the same magic. And we, too, can sing the world into being.

Thank you, Ringworm. Thank you, Fungus.

Dreaming a New Dream

When I started writing this book, my editor asked me: Where will this book lead to? How will it end? I couldn't give her an answer because I didn't know. I was going to journey to find out. And I can tell you, this adventure took me to places I never could have imagined beforehand. I'm deeply grateful and forever changed.

What have I learned? And where has it taken me?

First, I now understand that everything in nature is conscious, sentient, intelligent, and highly aware. Anyone can listen, hear, and understand the voices of nature. You just need to be willing to open yourself to them and *listen*. Nature *wants* us to know. Don't we all want to feel seen and heard for who we really are? That's what Nature wants, too.

Second, I am no longer annoyed! After the past year of journeying, when spring 2022 arrived (very slowly this year, which was a *little* annoying), I felt completely different about my garden. It was no longer about having a list of "to dos" that must be done or else I feel unhappy. I threw my own tyranny of tidiness out the window. I was able to relax and enjoy all the plants and animals that formerly annoyed me. Mugwort makes me happy wherever it pops up, and I ask permission to pick it to add to soups or bouquets or for smudging. (I even discovered that mugwort is a sacred plant to the Pennsylvania Dutch.) I thank it all the time for leading me down the path of love. I can enjoy the rabbits chasing each other around my yard (as long as they don't learn how to jump up onto my raised beds). And I am much more comfortable and accepting of the wildness creeping into my garden from the woods. I am more *gentle* and patient as I prune and clear away garden debris, leaving last year's leaf litter as mulch instead of bringing in mulch from somewhere else. And when I see a bug, I am much more likely to greet it with a hello rather than the bottom of my shoe. In fact, I noticed that because I didn't clean up the leaf litter beneath my perennials, I heard crickets more than a month earlier than I ever had before. (I love crickets and will *never* eat them.) And as the summer became drier and hotter, rather than trying to fix everything, I simply observed and witnessed the changes, making note of how *I* need to change for next year.

Now, when I find myself in an overly tidy, chemical-laden landscape I can feel the death. It's hard for me to even breathe.

Third, I am more convinced than ever that shamanic journeying is a powerful tool — at least for me. I have been convinced ever since an experience I had back in 2013, which I have kept secret up until now. Those were my earliest days of journeying — and during a journey, I was given the message to write children's books. And I was told the pen name that I should use when I wrote those books. I was very excited about this message, because I had already wanted to launch a children's book imprint. I was convinced that true health and happiness begins in childhood and can be epitomized by the love a child feels while sitting in a parent's lap and being read to. And parents can learn from reading to their kids as well. Rodale Kids books was launched in 2017. Two books in the launch were mine, released under my secret pseudonym. Yes, I am Mrs. Peanuckle! The name gifted to me during a shamanic journey. My ninth and tenth Mrs. Peanuckle alphabet books will be published in 2023 . . . The Earth Alphabet and Ocean Alphabet, two of my most favorite subjects. Journeying *works*.

Lastly, and perhaps most important, I have come to understand that our unique role in healing planet Earth is to heal ourselves. Healing ourselves means healing our hearts, and healing our hearts means learning how to love. Learning how to love means being kind, peaceful, and generous (and not doing stupid, hurtful things to others — even anonymously). What unites all of humanity — all the religions, all the countries, all races and ethnicities, all the genders and all the nature beings, is love. Plain and simple, love. *If we choose to practice it.* You have a choice. We all have a choice. I choose love. Thanks to journeying, I now know that we must heal the human heart in order to heal the Earth.

The concept of healing has been a lifelong mission for me. I literally grew up in the health and organic industry, and looking back on those sixty years, I can see that the industry got a lot of things wrong: the obsession with weight loss, fat, and food fads and fears, for example. But we also got some things right. Regenerative organic agriculture and environmental management is essential to all of us, and not just because we eat food — in fact, that's the *least* of the reasons why. More significant is that the needless poisoning of the Earth harms all of us — our bodies, our communities, our water, our soil, our air, our hearts, and our spirit — *because it also harms all the beings in nature that humans depend on for life.* The global environmental crisis is real. It is very likely that over the next decades, those of us still alive will witness unprecedented human migration. Vast numbers of people may be forced to flee their homes due to rising sea levels, desertification, starvation, lack of water, floods, and all the political unrest kindled by those sorts of tragedies. Instead of building walls, we need to figure out how to build better bridges (and boats). Instead of focusing on killing weeds, we need to plant more trees. Instead of dreaming of an unrealistic romanticized past where everything was "pure" (which it never really was), we need to dream a new dream where diversity is purely wonderful and appreciated as an important indicator of health and vitality. Instead of worshiping at the altars of success, fame, and wealth, we need to celebrate truth, kindness, and love. Instead of fearing there is not enough and feeling jealous of those who have more than us, we can choose to live by the rule of generosity — that through giving to others, we will discover there is enough for everyone. When we overcome our jealousies and focus on giving good things, even if it's just good thoughts, the benefit to all of us, but especially the individual who practices this (you), is exponential. Reciprocity and regeneration are the magic wands we each hold in our hands, and I deeply believe that by using our power for good, we can create a future that is not dominated by fear, war, and suffering.

Many apocalyptic religions and cultural myths include a belief in a judgment day or end-time, when everything falls apart and only the faithful are saved. It's a common tool used to scare people into obedience and appeals to our addiction to drama. But here is what I believe:

> *Every day is the end-time.*
> *Every day is judgment day.*
> *Every day is the birth of a new world.*
> *Every day is the second coming.*
> *Every day is everything.*
> *Act accordingly.*
> *Dream a new dream.*

Don't fret over every single thing you put in your mouth or how much you weigh. Go outside. Put your hands in the dirt. Grow something. Sit on a rock and listen to what nature is trying to tell you. Listen to your heart to learn what your purpose is. (Here's a clue: It will be connected to what creates the most joy in you.) Dance! (And don't make fun of people who do.) Feel the life everywhere around you — in these words, this book page, and the trees that made this paper. Or the energy that illuminates this page if you are reading on an electronic device. We are all physics! Nature is physics. Our bodies are physics. The quantum world is real, and it is magic. Feel the magical presence of all the beings — both human and otherwise — that exist on the web of life that have brought you to this very moment. Feel the love. Feel the gratitude. *Show* your gratitude. Be generous with everything. Sing! *We are amazing.*

These days, I read and hear so much about the loss of trust — in government, in the church, in the media, in the judicial system, in each other. And truth seems more subjective than ever. It can feel confusing and frightening. What is really true? Who can we really trust? What can we really trust? Will some people turn another horrific elementary school shooting into an alleged "false flag" operation? Will we all retreat back into our corners after every tragedy, letting television or social media tell us what to think? Will we let extremists threaten our freedom and steal our joy?

I hope not.

I say turn off your TV. Put down your phone. Here is what we can trust:

We can trust nature.

The sun. The moon. The seasons. The plants. The sassy animals. The bugs. (The ones we love *and* the ones that annoy us.) All the food and water that Earth provides. None of us can live without these things, and it behooves us to show our gratitude continually, and daily. We don't need fancy rituals. We don't need approval from any higher ups (although many of them are available for meetings in the Upper World if you'd like to speak to the Managers). We don't need permission. We can just go outside and give thanks. Acknowledge the warmth and light of the sun. Appreciate the phases of the moon that create waves and tides and cycles all over the Earth. Enjoy each season for what it brings — the fragrant exuberance of Spring, the delicious bounty of Summer, the savory harvest of Autumn, and the quiet rest of Winter.

Sometimes I wonder what will happen to my garden after I die. In less than a year the weeds would become predominant. Grass would grow tall, relax, and let itself run free. Thorny things would spread and create safe homes for animals and birds. Vines would climb up and curl around the fences and walls. The trees and plants will live and die and be born again as seedlings that will create new forests. Stones might be the last to disappear. A thousand years could pass in the blink of an eye and perhaps there would be mounds of earth made from fallen leaves, mosses, fungi, and ferns where once my magic garden lived. Or perhaps it will all be underwater like it was in some previous geologic era.

I live near many famous and beautiful historic gardens—Winterthur, Longwood, Chanticleer. Ironically, all were created through enormous wealth that came from family-owned chemical and pharmaceutical companies such as DuPont and Merck. And their beauty requires vast resources of money and people to maintain them. My garden is nothing like those flamboyant (but beautiful) places, and the last thing I want is to "preserve" my garden for posterity and create some sort of foundation to maintain it. That would be too much work. And yet gardening is certainly a LOT of work. So why do I keep doing it?

I do it because whenever there is extra space and time in my life, it's the first thing I *want* to do. I want to plant plants. I love watching young trees grow over time. I love to shop at nurseries and see what is new. I even like to spend time weeding because when I weed, I can study up close what's happening in my garden. I love creating serene views that I can enjoy from the windows of my kitchen and bedroom and writing studio. I love the smell of the earth on my hands and the flowers of every season. I love watching the chipmunks scamper, the groundhogs pretend to be invisible, the deer romping around and the birds flying this way and that. I love discovering the mysteries. And I eat, preserve, and share the delicious things I grow with family and friends. I have followed my joy and it has led me to my garden.

My garden is personal and intimate. I am an artist and I paint with plants and stone and soil and trees. We are collaborators, nature and me. The wildness is welcome because the wildness is me. But I am utterly delighted when other people recognize the magic — especially the young and old worker guys who come up here to fix or build things for me. Every single one of them, no matter what their political leanings, recognizes the magic and remarks on it. Their unsolicited reactions give me faith that I am on the right path.

What would make me, *and the Earth*, happiest is for this wildness to spread beyond my garden, beyond the Sand Pit, beyond the mountain, all across the country, the continent, the world, everywhere. For everyone to decide to stop adding toxins to the Earth and let it grow by being loved instead. Whatever space you live on can become a wildlife habitat, a nature sanctuary, a place of peace and harmony. The joy that arises from my garden belongs to everything and everyone — especially the plants, animals, birds, insects, reptiles, rocks, fungi, and microorganism that have thrived in this little Eden that will outlive all of us, even our children and their children.

My kids don't seem nearly as obsessed about gardening as I have been. I have given them the freedom to live their own lives and they have grasped it — floating like dandelion, thistle, or milkweed seeds on the wind, establishing their own roots, their own colonies. They are free to dream their own dream. But while we are all free in this moment to dream, we are free because of the dreams our ancestors dreamed for us. We are the dream our ancestors dreamed. Now, what will we dream for our grandchildren and great-grandchildren? What kind of seeds will we plant for the future we hope they will live in? What kind of gardens will we create to nurture the world we want to see?

Shamanic journeying cracked open and softened my heart in a way that I desperately needed and deeply appreciate. My hope is that in sharing these journeys, I can offer you a glimpse of what's possible. I hope that reading these stories helps you learn to trust and follow your own heart. Let yourself dream. We are all in the process of creating the future together. Let's make it amazing!

Trust love.

Trust nature.

Trust the magic.

CLOSING SACRED SPACE

(Maria style)

To the East, thank you for your guidance and protection.

To the South, thank you for your guidance and protection.

To the West, thank you for your guidance and protection.

To the North, thank you for your guidance and protection.

To the Earth, thank you for your guidance and protection.

To the Universe! Thank you for your guidance and protection.

Thank you!

GRATITUDE

First, this whole shamanic journeying thing would have never happened without Kathleen. I owe her a debt of gratitude for going first to the shaman. Kathleen, thank you for *everything*!

Then I need to thank Lisa Weikel, for being the shaman who helped heal my heart and guide me toward my new path. She also made sure everything I say regarding shamanism in this book is accurate. Deep, deep gratitude.

Here's to the brave early readers who encouraged me to share these stories and told me I was not 100 percent crazy: Bob Teufel, John Grogan, Judith Stiles, Jason Downs, Edwina Von Gal. Maya Rodale. What a wonderful and interesting crew. Thank you.

Thanks to Michael Jonn, the architect who told me about Ted Andrews.

Then there is my editor FERN! Who was a bit skeptical at first, but then saw the magic and believed in my irreverent approach. The fact that she was one of the editors of *Rodale's All-New Encyclopedia of Organic Gardening*, and then chose (along with everyone at Chelsea Green) to publish this book feels like completing a full spiral. I am thankful that I could trust her, and working with her has been a delight.

Thank you to my agent, Adriana Stimola for accompanying me on this long and winding road. I can't wait to see where it takes us! And thanks to her Mom, Rosemary Stimola, who represents Mrs. Peanuckle.

Thank you so much to everyone at Chelsea Green, especially Margo Baldwin, for welcoming me and my book into their independent, thoughtful, and environmentally responsible publishing company.

To the many helpers along the way: Michael Pollan, especially, for being the first to tell me I should write about journeying without drugs and for continuing to respond to all my emails and adding to my Trail of Books. From our lunch at Lutèce a thousand lifetimes ago, you have been an inspiration to me. Dr. Maya Shetreat for helping me shift my perspective on my writing. Jon Rassmussen for giving me healing guidance and being the first to tell me I wasn't going to be CEO for much longer (I didn't believe it at the time, but you were right). Bo Montenegro for teaching me at Esalen and in Sedona and letting me pass on the shaman's cave experience. Maria Lucia, who has passed into the great beyond. And Alison McGee, my Aussie sister from another lifetime. To Shawn Dove, who shares my belief in the power of love (and poetry) from when we were just kids. And Jonathan Leonardo, the preacher on the plane. And Bailey Torbert, whose golden light shines bright. The path of love is filled with wonderful helpers. I am grateful to all of them.

To all the guys who helped me with my landscape over the years. Especially Reds Bailey, Bundy, and Mike, who were with me from the very beginning. And Heath! Our crazy minds think alike, but Heath has the machines and knows how to make it real. John Panza — you, too, made it all happen! And all the team at Erwin Forrest who built my studio, thank you. Also thanks to Tim Delaney and team for putting in the high raised beds. The Glenn family, for hunting gently on the land. And Tom Bull and the Herbein's Garden Center for supplying my enormous appetite for plants for decades (but I wish you were all organic).

And then there is Elvin, the magical elf who helps me with the garden, the animals, and the mice. We make a good team. I'm so grateful to him.

To everyone at the Rodale Institute, especially Jeff Moyer, for keeping the flame burning.

To the Wildlands Conservancy, for keeping things wild (but please keep it organic, too!)

To Frank (the rabbit), for always making me laugh.

A big thank you to Twitter and all the amazing people I follow there. I have learned so much about the world through you ("translate tweet" is super cool, by the way). Scientists, researchers, political people, artists, activists, people of all colors and sexualities, humans. Honestly, I don't understand why the whole world isn't on it. Because Twitter connects the world. But please, be nice. I'd especially like to thank Patricia.

Thanks to Henry Louis Gates, Jr., for making *Finding Your Roots*, which taught me so much about all the small but pivotal moments of history that have made us who we all are.

Thanks go to my siblings: Heather, Heidi, and Anthony. It was hard, but we did it.

To Paul McGinley and Nicole Taylor, thank you for cleaning up our old corporate life so I could focus more on my new life. You have saved me.

Gratitude to my spiritual sister self-care team: Pam Fullerton, Freedom Flowers (yes, that's her real name), Holly Walck, and Sandy Stola. You have all kept me alive. Thank you.

To Lou, who is a great father and grandfather and fellow traveler on this path of life.

I have been blessed with wonderful friends who love me as I am and share the belief in magic with me. Kimbal Musk, you have helped me find my confidence, even when it's the confidence to disagree with you. And your work with Big Green helping kids learn to garden is an inspiration to all. David Totah, thank you for bringing me back to art and helping me understand my Jewish roots. I love you guys so much.

And the "Falcon," who loved me when I needed it most and let me go when I most needed to find freedom, even though it felt like death at the time. Trying to understand us was a major factor in me initially exploring shamanism. Thanks for giving me a reason to keep searching for answers. You made the right choice.

My kids are totally, totally awesome. Maya and Tony, Eve, and Lucia — I love you all so much. Even though you don't always understand my unusual hobbies, you still show up, eat my food and help me cook, laugh together with me, enjoy my magic garden and love me. And my eldest granddaughter, who once told her mother I have a "grocery store" in my backyard. And my newest granddaughter, who has yet to explore my magic garden and magic forest. Fun times ahead!

Every night as I rest my head on my pillow, I list all the things I am grateful for: the people in my life, the experiences, all the beings that surround me, my bed!

I'm grateful for you, dear readers. Each and every one of you. (Even the ones who are going to roll their eyes and think I'm crazy. Yup, you too.)

I am especially grateful to all the nature beings who kept annoying me until I realized they were trying to get through to me. And when I finally listened, they generously shared their insights, guidance, and love. You have changed me for the better forever.

And I'm grateful for my garden, where this book was born and written.

Dear reader, if you have gotten this far and are wondering what you can do, start where you are. What you are seeking starts within you. Find your own Trail of Books. Follow your curiosity and your joy. Sing. Dance. Dream. Learn to love nature. Trust the magic!

XO,
Maria

THE TRAIL OF BOOKS

I've listed the following books in their order of importance in my own journey. But if a particular book on the list is calling out to you, start there.

Andrews, Ted. *Nature-Speak.* Tennessee: Dragonhawk Publishing. 2004.
My first introduction to learning to communicate with nature.

Andrews, Ted. *Animal Speak.* Minnesota: Llewellyn Publications. 1993.
Great resource for learning and connecting with birds, animals, reptiles, and bugs.

Harner, Michael. *The Way of the Shaman.* San Francisco: Harper & Row. 1980.
First there was Carlos Castaneda (in the 1970s) to introduce the magic of shamanism to Americans. Then came Michael Harner, who brought academic rigor and credibility.

Harner, Michael. *Cave and Cosmos.* North Atlantic Books. 2013.
After a lifetime of studying shamanism, this is Harner's final report before his passing at age eighty-eight in 2018.

Ingerman, Sandra. *Shamanic Journeying: A Beginner's Guide.* Sounds True. 2003.

If you are truly interested in journeying on your own, this is a great book to start with.

Villoldo, Alberto. *The Four Insights: Wisdom, Power, and Grace of the Earthkeepers.* Hay House. 2006.

The founder of the Four Winds School of shamanism, Alberto Villoldo, who has a PhD in Psychology, has written many books, all of them interesting.

Villoldo, Alberto. *The Wisdom Wheel.* Hay House. 2022.

Villoldo's most recent book includes many helpful insights, including the importance of Anyi — or Reciprocity.

Weikel, Lisa. *Owl Medicine.* Xlibris. 2000.

Lisa's first and only book (to date!) tells the story of her early days as an aspiring shamanic practitioner.

Hawken, Paul. *The Magic of Findhorn.* Bantam Books. 1976.

Yes, THE Paul Hawken. His first book tells the story of the nature spirit infused commune in Scotland called Findhorn.

Wright, Machaelle. *Co-Creative Science.* Perelandra Ltd. 1997.

I found her books to be fascinating, a bit unbelievable and complicated, but very helpful in understanding what was happening to me in regard to communicating with nature.

Ruiz, Don Jose. *The Wisdom of the Shamans.* Hierophant Publishing. 2018.

I read this book after my book was done, but it further illuminated some of my journeys and is a helpful insight into the Toltec stories and traditions.

RESOURCES FOR GARDENERS

There are a zillion good books for gardeners; I can't begin to list them all. The resource I find indispensable and use the most are the apps called *Picture This*. There is one for plants, one for insects, and one for fungi. Just take a picture of what you want to identify; the app scans the photo and tells you what the plant, bug, or mushroom is and what it's good for.

Here are my favorite other sources for seeds, products, and information:

High Mowing Organic Seeds
    https://highmowingseeds.com

Peaceful Valley Farm and Garden Supply
    https://groworganic.com

Seed Savers Exchange
    https://www.seedsavers.org

Gardener's Supply Company
    https://gardeners.com
    (Wonderful offerings of gardening tools, equipment, and gadgets, and for raised bed kits.)

Alexis Nikole
    @blackforager on Instagram
    @alexisnicole on TikTok

RESOURCES FOR SHAMANIC STUDIES

The Foundation for Shamanic Studies
    https://shamanism.org

The Society for Shamanic Practice
    https://shamanicpractice.org

ABOUT THE AUTHOR

Paul Pearson

Maria Rodale is an explorer in search of the mysteries of the universe. Author, artist, activist, and recovering CEO, she serves on the board of the Rodale Institute and is also a former board co-chair. Throughout her career, she has advocated for the potential of organic regenerative farming to heal the damage wrought by pesticides and industrial agricultural practices. She is the author of *Organic Manifesto* and *Scratch* and is a secret children's book author. She was also featured in the documentary *Kiss the Ground*. Maria is a mother, grandmother, and crazy gardener who lives in Pennsylvania, right near where she was born.

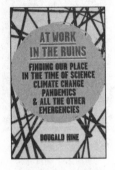

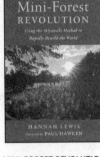

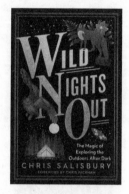